H. Furumai • S. Sato • M. Kamata
K. Yamamoto (Eds.)

AF184445

Advanced Monitoring and Numerical Analysis of Coastal Water and Urban Air Environment

Springer

Editors
Hiroaki Furumai
Professor
Research Center for Water Environment Technology
The University of Tokyo
7-3-1 Hongo, Bunkyo-ku
Tokyo 113-8656, Japan
furumai@env.t.u-tokyo.ac.jp

Shinji Sato
Professor
Department of Civil Engineering
The University of Tokyo
7-3-1 Hongo, Bunkyo-ku
Tokyo 113-8656, Japan
sato@coastal.t.u-tokyo.ac.jp

Motoyasu Kamata
Professor Emeritus
Department of Architecture
The University of Tokyo
7-3-1 Hongo, Bunkyo-ku
Tokyo 113-8656, Japan
and
Professor
Department of Architecture
Kanagawa University
3-27-1 Rokkakubashi
Kanagawa-ku, Yokohama
Kanagawa 221-8686, Japan
kamat@kanagawa-u.ac.jp

Kazuo Yamamoto
Professor
Environmental Science Center
The University of Tokyo
7-3-1 Hongo, Bunkyo-ku
Tokyo 113-8656, Japan
yamamoto@esc.u-tokyo.ac.jp

ISSN 1865-8504 e-ISSN 1865-8512
ISBN 978-4-431-54080-9 ISBN 978-4-431-99720-7 (eBook)
DOI: 10.1007/978-4-431-99720-7
Springer Tokyo Berlin Heidelberg New York

This work is subject to copyright. All rights are reserved, whether the whole or part of the material is concerned, specifically the rights of translation, reprinting, reuse of illustrations, recitation, broadcasting, reproduction on microfilm or in any other way, and storage in data banks.
The use of general descriptive names, registered names, trademarks, etc. in this publication does not imply, even in the absence of a specific statement, that such names are exempt from the relevant protective laws and regulations and therefore free for general use.

Cover Photo: Blue Tide in Tokyo Bay, Makuhari, Chiba Prefecture; © Masahiko Isobe

Printed on acid-free paper

Springer is part of Springer Science+Business Media (www.springer.com)

© 2010 to the complete printed work by Springer, except as noted. Individual authors or their assignees retain rights to their respective contributions; reproduced by permission.
Softcover reprint of the hardcover 1st edition 2010

Preface

Various environmental issues are related to urban activities. Through the growing recognition of the necessity to develop sustainable urban management, the University of Tokyo established the Center for Sustainable Urban Regeneration (cSUR) in July 2003. A research program at the cSUR was designed to create an integrated approach and to provide knowledge for sustainable urban regeneration with the aid of a global network of researchers and professionals, and to coordinate the international research alliance made up of leading academic institutions worldwide.

As part of the program, several studies have been conducted focusing on urban environmental problems in Asian megacities such as Tokyo, Taipei, Guangzhou, Shenzhen, and Bangkok. The following topics in particular were selected for integrated and strategic research supported by researchers from the fields of architecture, civil engineering, and environmental engineering:

–Integrated analysis of the urban atmospheric environment and its relationship with control of indoor air conditions in East and Southeast Asian countries

–Dynamic behavior of urban non-point pollutants in coastal environments

The research contains interesting intensive field-monitoring data on the coastal environment and the urban air environment. Topics also include state-of-the-art environmental monitoring and simulation analysis in urban areas. Key aspects of the research in advanced monitoring and the application of environmental numerical simulation were selected for inclusion in this book.

Integrating the monitoring and modeling of urban environments is essential for engineers to identify and investigate environmental problems and their solutions. In addition, advanced understanding of environmental phenomena is necessary to manage contemporary environmental issues. Environmental monitoring provides information about the processes and activities that characterize environmental quality. Model development cannot proceed without scientific data on environmental phenomena and the kinetics of associated processes. To understand the phenomena and processes, monitoring and modeling are fundamental.

The academic sector should update and add to the information on urban environments by discovering novel pollution phenomena and clarifying critical process mechanisms for pollution control. I hope that this book will be useful to undergraduate and graduate students and to experts and policymakers to improve their understanding of the field of environmental monitoring and model simulation.

Hiroaki Furumai

Contents

List of Contributors

Yoshihiko Akamine
Project Assistant Professor
Department of Architecture
Graduate School of Engineering
The University of Tokyo
Tokyo, Japan
akamine@arch.t.u-tokyo.ac.jp

Kensuke Fukushi
Associate Professor
Integrated Research System for
Sustainability Science (IR3S)
The University of Tokyo
Tokyo, Japan
fukushi@ir3s.u-tokyo.ac.jp

Hiroaki Furumai
Professor
Research Center for Water
Environment Technology
The University of Tokyo
Tokyo, Japan
furumai@env.t.u-tokyo.ac.jp

Hideaki Hoshino
Nihon Sekkei Inc.
Tokyo, Japan
hp_shellfish@hotmail.com

Masashi Imano
Assistant Professor
Department of Architecture
Graduate School of Engineering
The University of Tokyo
Tokyo, Japan
imano@arch.t.u-tokyo.ac.jp

Motoyasu Kamata
Professor Emeritus
Department of Architecture
The University of Tokyo
Tokyo, Japan
and
Professor
Department of Architecture
Kanagawa University
Kanagawa, Japan
kamat@kanagawa-u.ac.jp

Hiroyuki Katayama
Associate Professor
Department of Urban Engineering
The University of Tokyo
Tokyo, Japan
katayama@env.t.u-tokyo.ac.jp

Yukio Koibuchi
Associate Professor
Department of Socio-Cultural Environmental Studies
Division of Environmental Studies
Graduate School of Frontier Sciences
The University of Tokyo
Tokyo, Japan
koi@k.u-tokyo.ac.jp

Fumiyuki Nakajima
Associate Professor
Environmental Science Center (ESC)
The University of Tokyo
Tokyo, Japan
nakajima@esc.u-tokyo.ac.jp

Tassanee Prueksasit
Lecturer
Department of General Science
Chulalongkorn University
Bangkok, Thailand
Tassanee.C@Chula.ac.th

Shinji Sato
Professor
Department of Civil Engineering
The University of Tokyo
Tokyo, Japan
sato@civil.t.u-tokyo.ac.jp

Yu-Feng Tu
Cydea Incorporated
zacktu@gmail.com

Kazuo Yamamoto
Professor
Environmental Science Center (ESC)
The University of Tokyo
Tokyo, Japan
yamamoto@esc.u-tokyo.ac.jp

Yunchan Zheng
Nikken Act Design
Tokyo, Japan
zheng.yunchan@nikken.co.jp

1. Significance of Advanced Monitoring and Application of Environmental Numerical Simulation

Hiroaki Furumai

1.1 Introduction

In the fields of civil engineering, architecture, and environmental engineering, environmental monitoring and model simulation are essential component in setting up the strategy of environmental research from the aspect of urban sustainability. Advanced knowledge of monitoring and numerical simulation is required for graduate students to conduct their research dealing with the identification of environmental problems and investigations toward their solution. When dealing with increasing environmental concerns associated with water, air and soil pollution, as well as climate change induced by human activities, the accurate assessment of the state of the environment is a prerequisite for undertaking any course of action towards improvement.

In the twenty-first century COE research project, several studies have been conducted that focus on urban environmental problems in mega cities in Asian countries such as Tokyo, Taipei, and Bangkok. The research works contain interesting and intensive field monitoring data on coastal environmental and urban air environments. Topics also included state-of-the-art of environmental monitoring and simulation analyses in urban areas: of hazardous substances, atmospheric movement, coastal hydrology, biological tests, and wastewater.

Before the representation of the COE research projects, this chapter will first discuss the significance of environmental monitoring and numerical simulations. Environmental monitoring provides information to describe the processes and activities, which characterize the environmental quality. Monitoring data is used in the preparation of environmental impact assessments, as well as in many circumstances in which urban activities carry a risk of adverse effects on the natural environment. In addition, monitoring

H. Furumai et al. (eds.), *Advanced Monitoring and Numerical Analysis of Coastal Water and Urban Air Environment*,

© 2010 to the complete printed work by Springer, except as noted. Individual authors or their assignees retain rights to their respective contributions; reproduced by permission.

data is required to construct environmental models and to calibrate and validate models.

Environmental models seek to reproduce that occurs in a certain area during certain events. It is much easier and more practical to create mathematical models and run certain experiments than to go out and do the same experiment in an actual environment. All models have a specific target area and should be developed in accordance with their purpose of modeling. This means that it is crucial to determine the target area and to formulate the target phenomena with involved processes. In addition, the purpose of modeling should be clearly defined, considering the required accuracy of reproduction by the model.

Model development cannot proceed without scientific information and knowledge on environmental phenomena and the kinetics of associated processes. Monitoring and experiment are fundamental steps to understand these phenomena and processes. Such as planning and strategy must be well-designed in order to establish the current status of the environment and to understand trends in environmental quality parameters. The basic steps of and the cyclic process for model development and simulation are illustrated in Fig. 1-1.

Models have three basic parts, starting with the science, moving towards a mathematical representation of that science, and ending in the solution of

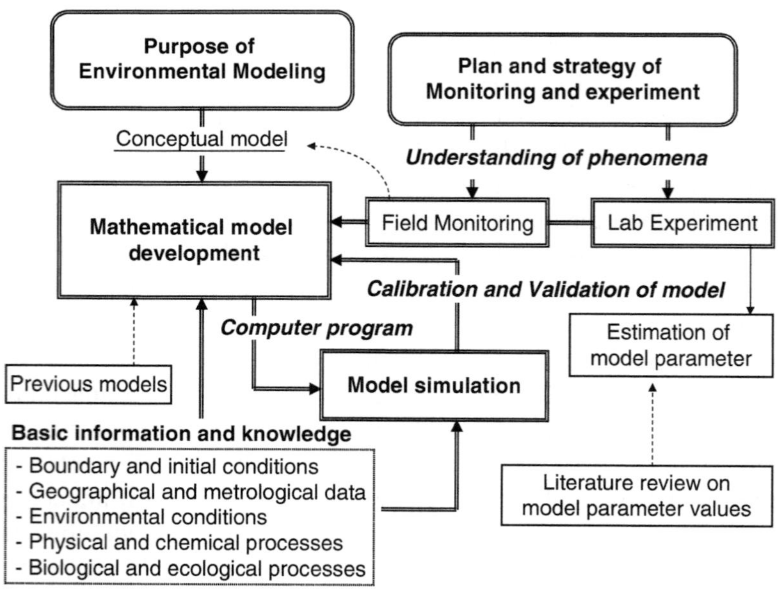

Fig. 1-1. Basic steps of and cyclic process for model development and simulation

the mathematics as a simulation. The overall process is actually a cyclical system: the answer that emerges from the simulation is used to refine the science, which leads to a new set of mathematics, which is expressed by a new computer program. One of the tasks we face in modeling is deciding when we think the answer is "right" enough.

Before the model application, the environmental models need to be calibrated and validated using monitoring and experimental data. Then the model can output the simulation results with given boundaries and initial conditions. A well-validated model can be used for future predictions within different scenarios, which are given considering future urban environmental planning. The simulation results are usually represented into a graphic form so as to depict their essential meanings. It is notable that graphic expression is also an important process in the entire of simulation research.

1.2 Water Quality Monitoring and Simulation for Sustainable Water Management

Since the author's specialty is water environment technology, the significance of monitoring and model simulation will be discussed with sustainable urban water management as an example. Many possible pathways to sustainable water management should be considered and relevant factors should be interrelated from the various aspects.

In order to explore sustainable water management, it would be necessary to evaluate and diagnose the water environment including water flow and quality. It is essential to know how to conduct the monitoring and how to construct models which can contribute to pollution control and management.

1.2.1 Evaluation of Water Cycle and Water Environment

First of all, we have to recognize the water cycle and water balance in the target watershed, since any water environment is deeply dependent on the characteristics of its related watershed. Rainwater falls on the ground and is stored in forests, soil, and groundwater. It then flows through rivers down to coastal waters and evaporates from lakes and seas to return as rainfall. This is the natural hydrological water cycle. The evaporation process contributes to the recovery of water quality, while the precipitation in mountainous areas provides the gravitational potential energy of water.

In urban regions, artificial water flow in water supplies and sewer systems coexists with the natural water flow as shown in Fig. 1-2 (Tambo 1976).

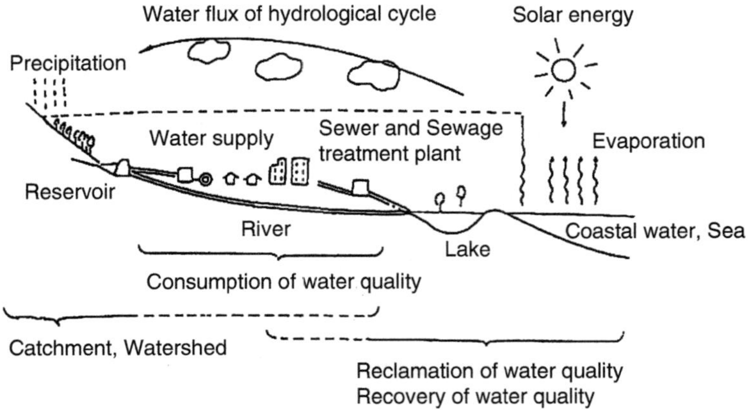

Fig. 1-2. Water environment and urban water systems in the entire water cycle

Dams and reservoirs have been artificially constructed to secure stable water resources. After the introduced water has been used in urban areas, the resulting wastewater or treated wastewater is discharged to receiving water. Urban water use has a high potential to impact the natural water cycle, even while we reap the benefits of various water usages. This impact can bring quite a lot of change in the water flow, in water quality and in the biological community which consists of the aquatic environment.

In order to understand the changes brought about by this impacts, it is necessary to quantify environmental components through monitoring. Additionally, it is desirable to evaluate them quantitatively using calibrated models. Figure 1-3 shows the important components of water environment and the interrelationships between the components and beneficial water use. Although the Figure is not sufficient to integrate all factors of influence, the afore-mentioned concepts and viewpoints are included. First, we have to evaluate the quality characteristics of available water resources and to discuss acceptable water use with appropriate management. In addition, we also must pay attention to the ecological impact made on the biological community as well as on the water flow and chemical quality.

Temporal and spatial changes in water resources and the water environment should both be considered in order to establish a stable water supply and to support safe water use. Since the required water quality depends on the types of beneficial water use, water resource distribution should be designed and planned in consideration of quantity, quality and their seasonal change. In other words, water resources and the environment should be monitored and modeled in consideration of water demands.

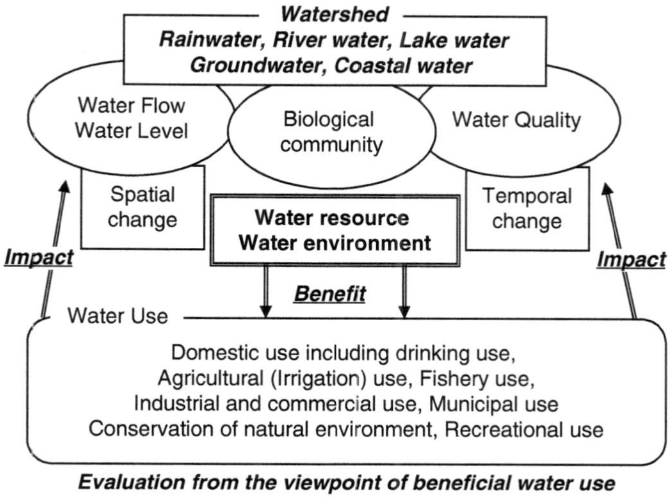

Fig. 1-3. Water environment components and beneficial water use

1.2.2 Environmental Quality Standard and Monitoring for Pollution Control

For water pollution control, environmental quality standards (EQS) are established as target levels for water quality that are to be achieved and maintained in public water bodies under the Basic Environment Law. The standards have two major goals: protection of human health and conservation of the living environment (Okada and Peterson 2000). The second goal is set for classified water bodies such as rivers, lakes, reservoirs, and coastal waters. The standard values for the living environment have been established for biochemical oxygen demand (BOD), chemical oxygen demand (COD), dissolved oxygen (DO), and other variables based on water usage. Table 1-1 lists the classified standard values related to the conservation of the living environment in rivers (Ministry of the Environment 1997). Each class corresponds to type of water use. Therefore, both the water quality of water resources is considered as well as their amounts. In this sense, the rapid growth of the urban population raises serious concerns about water availability from the two viewpoints of the rapid demand for high quality water and of water pollution after water use in urban area.

Water quality monitoring has been officially conducted and the monitoring data is commonly used for evaluating the water environment. Figure 1-4 shows the changes in the attainment rate in terms of BOD and COD as the

Table 1-1. Ambient water quality standards for river related to the conservation of the living environment

Rivers class	Item	Standard value				
	Water use	*p*H	BOD	SS	DO	Total Coliform
AA	Water supply class 1, conservation of natural environment and uses listed in A–E	6.5–8.5	1 mg l^{-1} or less	25 mg l^{-1} or less	7.5 mg l^{-1} or more	50 MPN/100 ml or less
A	Water supply class 2, fishery class 1, bathing and uses listed in B–E	6.5–8.5	2 mg l^{-1} or less	25 mg l^{-1} or less	7.5 mg l^{-1} or more	1,000 MPN/100 ml or less
B	Water supply class 3, fishery class 2. and uses listed in C–E	6.5–8.5	3 mg l^{-1} or less	25 mg l^{-1} or less	5 mg l^{-1} or more	5,000 MPN/100 ml or less
C	Fishery class 3, Industrial water class 1 and uses listed in D–E	6.5–8.5	5 mg l^{-1} or less	50 mg l^{-1} or less	5 mg l^{-1} or more	–
D	Industrial water class 2, agricultural water and uses listed in E	6.0–8.5	8 mg l^{-1} or less	100 mg l^{-1} or less	2 mg l^{-1} or more	–
E	Industry water class 3, and conservation of environment	6.0–8.5	10 mg l^{-1} or less	Floating matter such as garbage should not be observed	2 mg l^{-1} or more	–

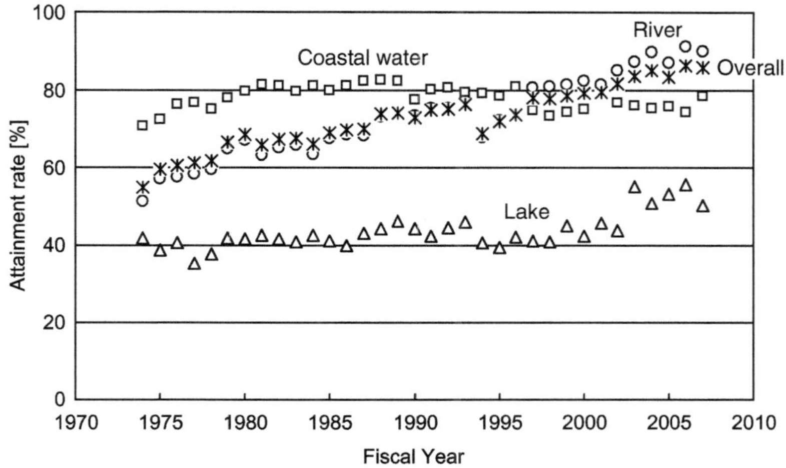

Fig. 1-4. Change in the attainment rate of water quality standards in public waters (River: BOD, Lake and Coastal water: COD)

organic pollution indicators since the year in which data was first collected on water quality management by the use of EQS. During the two decades of high economic growth from 1955 to 1975, the rapid spread of pollution in rivers and other water bodies was clearly evident in urban areas. After this period, river water quality has been improved around the country thanks to effective effluent regulation and the construction of sewerage systems with wastewater treatment plants. While the attainment rate of BOD in rivers was increased by mitigating organic pollution, the level of attainment of COD for water pollution remains low in enclosed water bodies such as bays, inland seas, and lakes with major pollution sources in their surrounding regions.

Efforts must be focused on effectively reducing pollutant loads in populated and industrialized areas around enclosed water bodies to improve their water quality. Additionally, specific regions contributing to water pollution are identified within each specified body of water. Every five years, the Minister of the Environment sets target on COD pollutant load reduction for each specified water area, as well as the target year by which they are to be met. These regulations, in accordance with the total pollutant load control standards, are the core of load-reducing measures based on the Areawide Total Pollutant Load Reduction Plan. These pollution control countermeasures have been implemented based on water quality monitoring data and water quality predictions by models. Combinations of monitoring data and model predictions are essential approach to manage water quality control effectively.

1.2.3 Significance of Advanced Monitoring and Modeling Research

Official water quality monitoring is designed so that the government can accurately evaluate the effectiveness of current countermeasures for pollution control. If there are no improvements in water quality, the government can deliberate the amendments as necessary and work to improve control measures. At the same time, the government also has to try to monitor any defects in the water environment as well as of water use impairment.

As mentioned in the last section, enclosed water bodies continue to suffer from pollution and are in need of long-term countermeasures. In order to maintain, conserve, and improve the functioning of lakes and coastal waters area, it is vital to ascertain and assess not only water quality, but also the aquatic environment of the area as a whole, including fish, bottom sediment, and benthic organisms. From this perspective, academic and administrative sectors have worked together to establish well-qualified methods to assess the water purification and other functions of enclosed water bodies, and quantitatively assess the aquatic environment as a whole.

Academic sectors must play a role in discovering newly relevant environmental phenomena and clarifying the pollution mechanism. It means that we have to conduct advanced environmental monitoring as well as officially monitored environmental indicators. Such advanced monitoring can provide new ideas and hints to capture unknown phenomena and processes. Then, a conceptual model can be built to express these mechanisms in the target water environment. This research process is the first step in developing a mathematical model, as shown in Fig. 1-1.

1.3 Monitoring and Modeling for Management and Research

1.3.1 Necessity of Water Environment Information on a Watershed Basis

The first version of the Basic Environment Plan was drawn up in December 1994 based on the Basic Environment Law, which outlines the general direction of Japan's environmental policies. The Basic Environment Plan is designed to engage all sectors of society in a concerted effort to protect the environment. The Plan maps out the basic approach of environmental policies with the mid-twenty-first century in view, and identifies four long-term objectives.

The Basic Environment Plan places "Conserving the Water Environment" among those policies under which we need to build a socioeconomic system which fosters a sound material/resource cycle. This system is closely coordinated with sustainable urban regeneration. The policy is aimed at achieving the integrated conservation of water quality and quantity, aquatic biota, and near-shore areas in order to maintain and enhance sound water cycles. Effective planning needs to be developed in order to solve various water problems and secure sound water cycles. To promote the kind of control and management, it is important to diagnose and review water balance and water pollution on the basis of watershed and to formulate a master-plan for securing water use based on sound water cycles in the watershed. It means that information on water environments should be integrated on a watershed basis, in consideration of urban activities that include water use.

Many authorities and sectors are involved in water use and water quality control in watersheds. Usually, river bureaus and administrators play the most important role in river flow monitoring and flow management for flood control. Water supply and sewerage services are also closely involved with water abstraction and water pollution control respectively. These bodies continue to monitor and record information on water quality of drinking water source and effluent. Environmental agencies and local environmental protection divisions are responsible for the control of effluent from factories and specified facilities, the conservation of water environment, and water quality monitoring in public water bodies. Agricultural policy planning departments can contribute to water balance and water quality management through the intake control of agricultural water and the proper management of fertilizers and pesticides applications.

Although these authorities and sectors give thought to the integration of information on water environments, they tend to be cautious of exceeding their respective authority and are self-restricting in terms of making ambitious policy on watershed management. The academic sector should contribute to the establishment of an efficient monitoring system and the development of an information platform for the integrated management of water environments.

1.3.2 Coordination of Monitoring and Modeling

In order to maintain and utilize any information platforms established, work on "Monitoring" and "Modeling" should be carried out in close coordination. Monitoring plans should be designed to calibrate the related model and enhance its function. Quantitative analysis and assessment can be undertaken by model-based simulation, with reference to the basic information/data contained on the platform. At the same time, any data and information required can then be identified through the model development process. A better and more quantitative understanding of water cycles and water quality changes is essential to the development of effective watershed management.

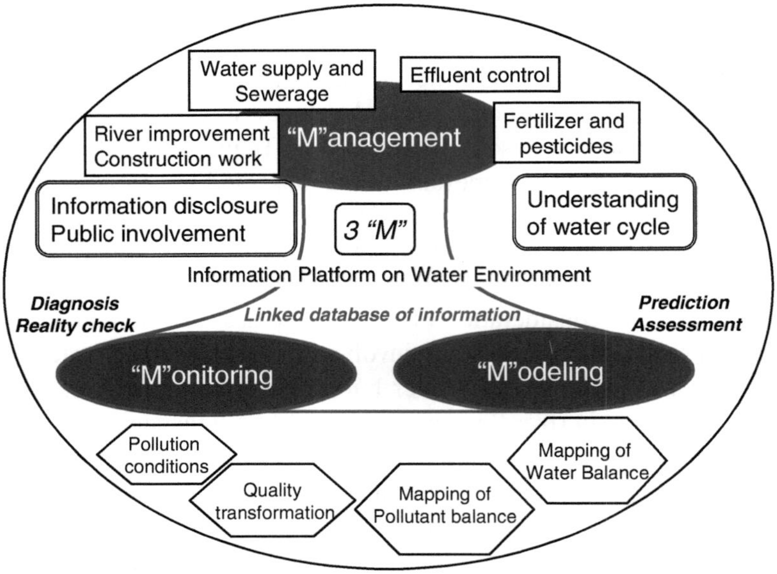

Fig. 1-5. Water management based on coordinated monitoring and modeling

Based on the outputs from two (Monitoring and Modeling), a third M, Management, must then function efficiently in various fields such as water supply and sewerage works, effluent and pollutant source control, river flow and dam control, agriculture and forest management. As shown in Fig. 1-5, these three Ms – Monitoring, Modeling and Management – serve as the mechanism for sustainable water use on the information platform on watershed environment.

The academic sector, such as universities, must contribute to the renovation and improvement of the information platform on water environment by discovering novel water pollution phenomena and clarifying critical process mechanism for pollution control. For this purpose, advanced monitoring should be considered from the viewpoints of high frequency and detailed spatial resolution. Intensive monitoring can provide new findings and evidence to support hypotheses, if they are well designed and planned. In addition, we must also pay attention to emerging micro-pollutants such as pharmaceutical compounds and newly threatening pathogens, which are not listed in the official parameters for monitoring. The application of new and original sensors and the analysis of multiple environmental parameters may well lead to discovery and new hypotheses for the formation of environmental phenomena.

Based on the results of advanced environmental monitoring, we can modify and upgrade existing simulation models for water management, which need to be supported by strong scientific evidence and quantitative information. Together with well-summarized and integrated knowledge, model-based prediction can improve the information platform on water environments, which in turn contributes to the proposal of effective strategy and the development of innovative management for sustainable water environments.

References

Ministry of the Environment (1997) Environmental quality standards for water pollution. http://www.env.go.jp/en/lar/regulation/wp.html

Okada M, Peterson SA (eds) (2000) Water pollution control policy and management: the Japanese experience. Gyosei, Tokyo

Tambo N (1976) Structure and capacities of water metabolic systems. J JWWA 497:16–34 (in Japanese)

2. Environmental Monitoring in Urban Coastal Zone

Fumiyuki Nakajima, Hiroyuki Katayama, Hiroaki Furumai, and Yukio Koibuchi

2.1 Urban Non-point Pollutants in Coastal Environment

2.1.1 Significance of Urban Non-point Pollution in Coastal Environment

Coastal water receives a large number of pollutants from urban area which have a potentially adverse effect on water quality and aquatic ecosystems. The sources of the discharge of such pollutants are categorized into two types: point and non-point sources.

Point source means pollutant sources which are easily identified facilities discharging pollutants to the environment. Factories, housings and sewage treatment plants are included in this category. In many countries, the quality of the discharged water from these point sources is regulated by laws and is controlled by a wide range of treatment technologies.

Diffuse sources of pollutants such as vehicles and urban surface materials are called non-point sources. Non-point source pollution occurs with rainfall or snowmelt. The water from rain or snow dissolves the atmospheric pollutants, washes off the pollutants on the impervious surfaces and finally flows into rivers, lakes and coastal waters. Naturally, the non-point source pollution is difficult to control, since the water is irregularly discharged. The monitoring or sampling of such irregular water discharge requires special devices and/or incurs high labor costs. The source responsible for the pollution is often unclear, not least because the water runoff itself is a natural phenomenon and the pollutant sources are diverse; responsibility is thus difficult to assign. As such, the significance of non-point source pollution in the water environment tends only to be recognized after the controlling system of the point sources has been spread well in the society.

H. Furumai et al. (eds.), *Advanced Monitoring and Numerical Analysis of Coastal Water and Urban Air Environment*,
© 2010 to the complete printed work by Springer, except as noted. Individual authors or their assignees retain rights to their respective contributions; reproduced by permission.

Urban runoff contains a variety of chemicals (Fig. 2-1). Nutrients, organic matter and suspended solids, which are most common concern in enclosed water bodies, are obviously comprised of urban runoff pollutants. Trace toxic compounds are generally found in urban runoff. Traffic activities produce a large number of pollutants as gas exhaust and particulates, tire/road surface abrasion, oil spill, and so on. Paints, surface materials and other chemicals used outside, like pesticides, are also washed off under certain conditions. The prioritization of these chemicals in urban runoff is still under discussion (Eriksson et al. 2005).

The real impact of urban runoff to the surrounding water environment is dependent on the drainage facilities in the area. Separate sewer systems bring wastewater from point sources to treatment facilities and drain the runoff water into the environment. Combined sewer systems convey both types of water to the treatment facility. However, excess water resulting from heavy rain causes the direct discharge of a mixture of untreated wastewater and runoff water. The importance of non-point source pollution control should be judged in consideration of the current situation of point source management and the type of sewer system in the catchment of concern.

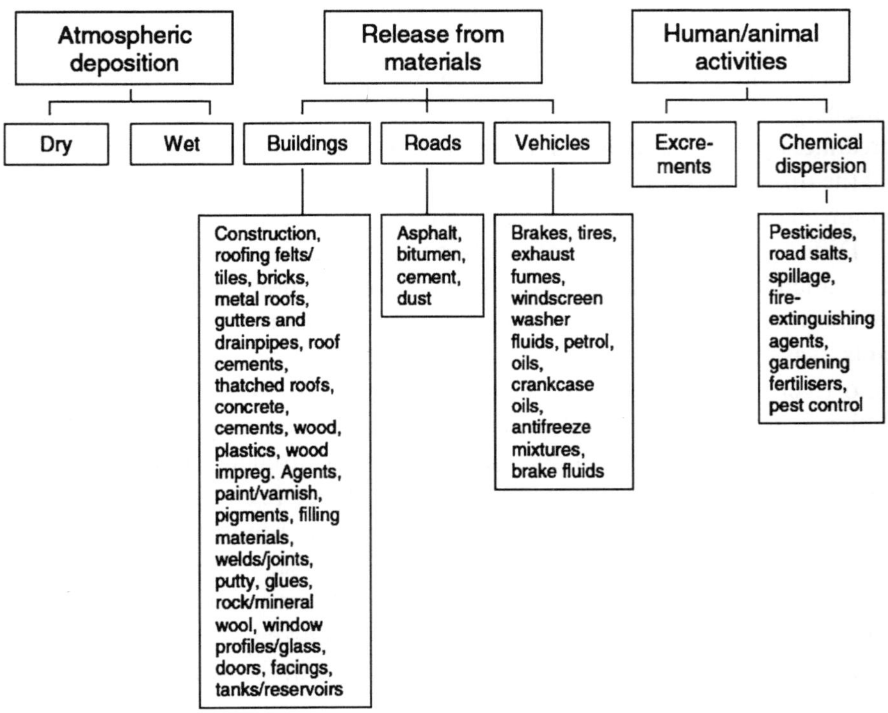

Fig. 2-1. Pollutants potentially found in urban runoff (Eriksson et al. 2005)

2.1.2 Polycyclic Aromatic Hydrocarbons as a Group of Urban Micro-pollutants and Their Sources

Polycyclic aromatic hydrocarbons (PAHs) are known to be acutely toxic, genotoxic, and carcinogenic compounds (Phillips 1983; Hagris et al. 1984; Baumann 1998). Micro-pollutants such as PAHs and heavy metals are widely distributed in dust, soils, and sediments, and are found in roof and road runoff (Hoffman et al. 1984; Takada et al. 1990; Sansalone and Buchberger 1997; Roger et al. 1998; Heaney et al. 1999; Förster 1999; Krein and Schorer 2000; Chebbo et al. 2001;Furumai et al. 2002; Brenner et al. 2002; Murakami et al. 2003). Hoffman et al. (1984) estimated that 36% of environmental PAH input was due to urban runoff; for the higher molecular weight PAHs, the figure was 71%. Urban runoff has been recognized as an important PAH pathway to water environments and aquatic ecosystems. Hence, effective control strategies for PAHs in urban runoff are required to ensure human and ecosystem safety.

Road dust has been recognized as bringing a large volume of PAHs into the water environment via road runoff (Brown et al. 1985; Maltby et al. 1995a,b; Boxall and Maltby 1995). Possible PAH sources in road dust include diesel vehicle exhaust, gasoline vehicle exhaust, tire, pavement (asphalt or bitumen), and oil spill. Based on the enrichment factor, Takada et al. (1990) identified vehicle exhaust as the primary PAH contributor to road dust collected from roads with heavy traffic, while atmospheric fallout was more significant in residential areas in Tokyo. Zakaria et al. (2002) suggested used crankcase oil as one of the major PAH contributors to road dust in Malaysia.

We estimated the comparative contribution from potential PAH sources in road dust samples, using a statistical approach based on a large number of reported PAH profiles (Pengchai et al. 2004, 2005). Seven kinds of PAH sources were defined: diesel vehicle exhaust, gasoline vehicle exhaust, tires, asphalt-pavement, asphalt or bitumen, petroleum products excluding tires and asphalt, and combustion products except those in vehicle engines. Using cluster analysis combined with principal component analysis, the 189 source data were classified into 11 source groups based on the content percentage of 12 individual PAHs (12-PAH profiles) (Table 2-1). Thirty-seven dust samples on nine streets in Tokyo were collected and subjected to PAH analysis both with and without particle size fractionation. Multiple regression analysis was applied to estimate the sources of the PAHs in the dust samples. The result demonstrated that the abrasion of tires and asphalt-pavement contributed a certain amount of PAHs to road dust, in addition to diesel vehicle exhaust, which has been recognized as the main source of PAHs in road dust.

Particle size and density are important parameters in wash off processes for urban runoff. PAH distribution in harbor sediment fractions has been

Table 2.1 PAH compositions of source materials (Pengchai et al. 2005)

	S1	S2	S3	S4	S5	S6	S7	S8	S9	S10	S11
Average PAHs profiles											
phenanthrene (Ph)	6	**26**	7	3	4	**31**	*16*	*17*	*11*	1	**48**
anthracene (An)	1	4	**43**	**27**	7	8	3	2	3	2	*15*
fluoranthene (Fr)	*13*	*22*	9	*15*	3	1	6	6	**51**	0	8
pyrene (Py)	**43**	**31**	8	*13*	*14*	2	*11*	*19*	6	0	*10*
benzo(a)anthracene (Ba)	0	4	3	**26**	*11*	7	5	2	4	1	5
chrysene (Ch)	3	4	3	1	*10*	*10*	*12*	2	4	1	2
benzo(k)fluoranthene + benzo(b)fluoranthene (Bf)	3	2	8	4	**28**	**27**	*15*	*10*	7	6	3
benzo(a)pyrene (Bpy)	4	2	4	3	0	0	8	**31**	5	*11*	3
indeno(1,2,3-cd)pyrene (In)	6	2	7	2	1	2	6	1	3	*23*	1
dibenz(a,h)anthracene (Db)	1	1	4	4	*21*	7	3	8	0	**46**	2
benzo(ghi)perylene (Bpe)	*19*	2	4	2	1	5	*15*	1	5	9	2
Total %	100	100	100	100	100	100	100	100	100	100	100
Sample number belonging to each group											
diesel vehicle exhaust [D]	1	*49*	*5*	–	–	2	2	–	–	–	20
gasoline vehicle exhaust [G]	–	8	2	*10*	*8*	1	2	–	–	1	21
tire [T]	*8*	–	–	–	–	–	–	–	–	–	–
asphalt-pavement [P]	–	2	–	–	–	–	*6*	–	–	–	–
asphalt or bitumen [A]	–	–	–	–	–	–	*9*	–	–	–	2
petroleum products excluding tire and asphalt [O]	–	1	–	–	–	–	–	–	–	2	7
combustion products except for those in vehicle engines [E]	–	1	1	–	–	–	–	*7*	*5*	–	6
Total number of sample	9	61	8	10	8	3	19	7	5	3	56

*Bold and italic values are characteristic PAHs species/source categories in each group

reported in both size and density (Ghosh et al. 2000; Rockne et al. 2002; Ghosh et al. 2003). Rockne et al. (2002) revealed that 85% of the total PAHs in Piles Creek sediment was found in the light fractions (<1.7 g/cm^3), despite the fact that light density components comprised only 4% of the total sediment mass. In addition, they suggested that the preferential sequestration in the Piles Creek sediment was likely due to the presence of detrital plant debris. Ghosh et al. (2000) showed that the coal/wood-derived particles (specific gravity < 1.8) constituted only 5% of Milwaukee Harbor sediment by weight, but contained 62% of the total PAHs.

We reported PAH concentrations in size- and density-fractionated road dust collected in Japan (Fig. 2-2) (Murakami et al. 2005). The percentage contribution by weight of light density particles to the total deposition mass was less, but the light fractions accounted for a significantly higher ratio of PAH mass in road dust. It is suggested that light fractions in road dust contribute significantly to stormwater contamination despite their minor contribution to the total deposition mass, due to their high PAH contents as well as their physical property (easily conveyed by surface runoff water). We also

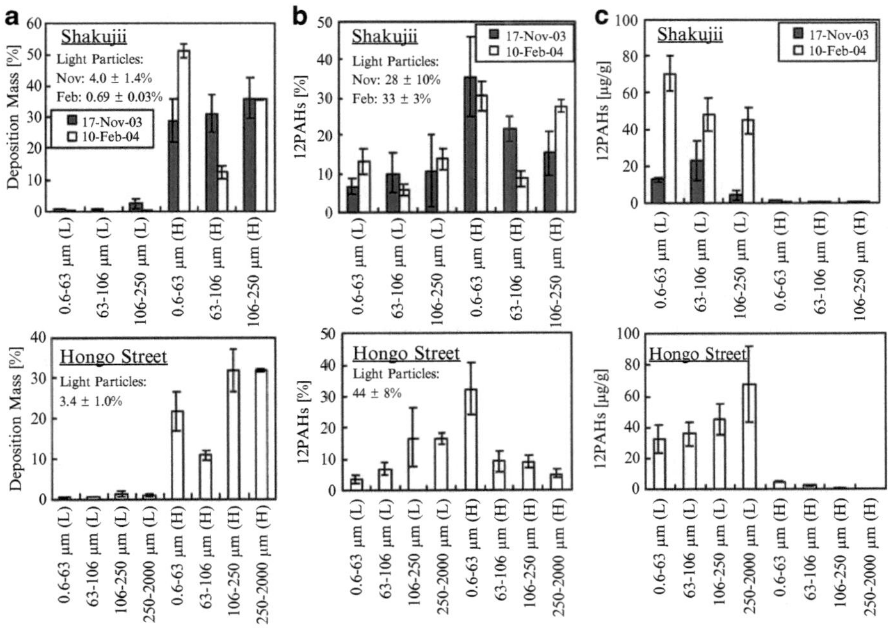

Fig. 2-2. Mass distribution (**a**), total 12PAH distribution (**b**) and total 12PAH content (**c**) in road dust by size- and density-fractions [mean ± S.E.] (*L* light particles, *H* heavy particles) (Murakami et al. 2005)

conducted a cluster analysis to reveal that there was a significant difference in the PAH profiles between locations rather than between size-fractions, density-fractions and period of sampling. Apart from the locations, the PAH sources might differ due to sampling time or size fractions. Multiple regression analysis indicated that asphalt/pavement was the major source of road dust in residential areas, and that tires and diesel vehicle exhaust were the major source of road dust in heavily trafficked area.

2.1.3 Monitoring and Modeling of Urban Runoff Pollutants

The monitoring of urban runoff often requires highly sophisticated equipment and human organizations to catch rainfall events whenever they happen. Runoff water quality varies with rainfall patterns and antecedent pollutant deposition conditions. Due to the difficulties inherent in regular monitoring of urban runoff, mathematical models are utilized to simulate rainwater runoff and pollutant transportation. Such models are useful in evaluating the effectiveness of pollution-control measures in protecting the water environment from non-point source pollution.

Runoff models for suspended solids (SS) have been developed by many researchers (Sartor and Boyd 1972; Tomonvic and Makishimovic 1996; Furumai et al. 2001; Hijioka et al. 2001; Uchimura et al. 1997). These models can be utilized in the simulation of particle-associated pollutants such as polycyclic aromatic hydrocarbons (PAHs). The runoff of particle-associated micro-pollutants is thought to depend on particle size distribution. Several field surveys have shown that micro-pollutants are attached to fine sediments and particles (Sansalone and Buchberger 1997; Roger et al. 1998; Murakami et al. 2005; Sartor and Boyd 1972). It has been reported that the runoff and sedimentation characteristics of fine particles are different from those in the coarse fraction (Furumai et al. 2002; Brenner et al. 2002; Roger et al. 1998; Tomonvic and Makishimovic 1996; Andral et al. 1999). Therefore, the SS runoff models need a modification of particle categorization to extend their application to PAH runoff models. Urban surface category is also an important factor in runoff modeling. Hijioka et al. (2001) proposed a SS runoff model with two particle size categories and two urban impervious surface types: fine (smaller than 45 μm) and coarse (larger than 45 μm) particles on roads and roofs. In the model, roof runoff was characterized as a faster process than road runoff because roofs have steeper slopes and smoother surfaces. As shown in the previous section, the PAH composition varies according to emission source. Murakami et al. (2003) revealed differences in PAH compositions between roof dust and road dust. The result of cluster analysis on PAHs profiles in the size-fractionated dust showed that the roof dust formed a separate cluster to the road dust, irrespective of either particle size or roof structure. Factor analysis revealed that phenanthrene, indeno(1,2,3-cd) pyrene and benzo(ghi)perylene were important PAHs for distinguishing the road dust and the roof dust. The result of the factor analysis also suggested that the contribution of tires, pavements or asphalts to PAHs was greater in road dust than in roof dust, and that the contribution of vehicle exhaust emission to PAHs was greater in roof dust than in road dust. A non-parametric test indicated that the PAHs content was higher in the fine dust (smaller than 106 μm) than in the coarse dust (larger than 106 μm).

We also developed a model explaining the dynamic runoff behavior of particle-associated PAHs (Murakami et al. 2004), in which roads and roofs were considered separately as impervious surfaces, and particle sizes were classified into fine and coarse fractions (Fig. 2-3). A field survey for model development was conducted in a densely populated area in Japan. Consideration of two types of road dust with different mobility is conceptually useful to explain the PAH profiles in runoff particles. Such a model scheme achieves good agreement with observations of SS and PAH runoff behavior for fine particles, except during heavy rainfall. To improve the disagreement, it may be necessary to take account of additional sources of SS and PAHs washed off by heavy rainfall.

2.1.4 Toxicity Evaluation of Micro-pollutants in Aquatic Sediments

Hydrophobic organic pollutants in urban runoff have the potential to accumulate in the sediments of receiving waters and may have adverse effects on ecological systems, especially on benthic organisms (Maltby et al. 1995a,b; Boxall and Maltby 1997). Biological toxicity tests are common tools for assessing the overall toxicity of contaminated sediments. Chemical analysis, usually including a process of organic solvent extraction, provides basic and quantitative information on the compounds of concern. However, the procedure does not reflect the biological process of ingestion, or the bioavailability of the compounds. To evaluate the toxicity of a specific compound in the sediment, mobility of the compound to biota, or bioaccessibility, is a key factor. Generally speaking, the dissolution of such sediment-associated hydrophobic contaminants is not well predicted from a simple partitioning model. The contaminants are not uniformly bound to the sediment: some are easily desorbable and others strongly bound to the solids. This difference of mobility in the toxic substances results in different bioavailability to the benthic organisms. Conventional evaluations of hydrophobic organic pollutants in sediments are based on the total amount of the pollutants, as determined by extraction using organic solvents. This, however, is not sufficient for understanding their effect on ecological systems (Kelsey and Alexander 1997).

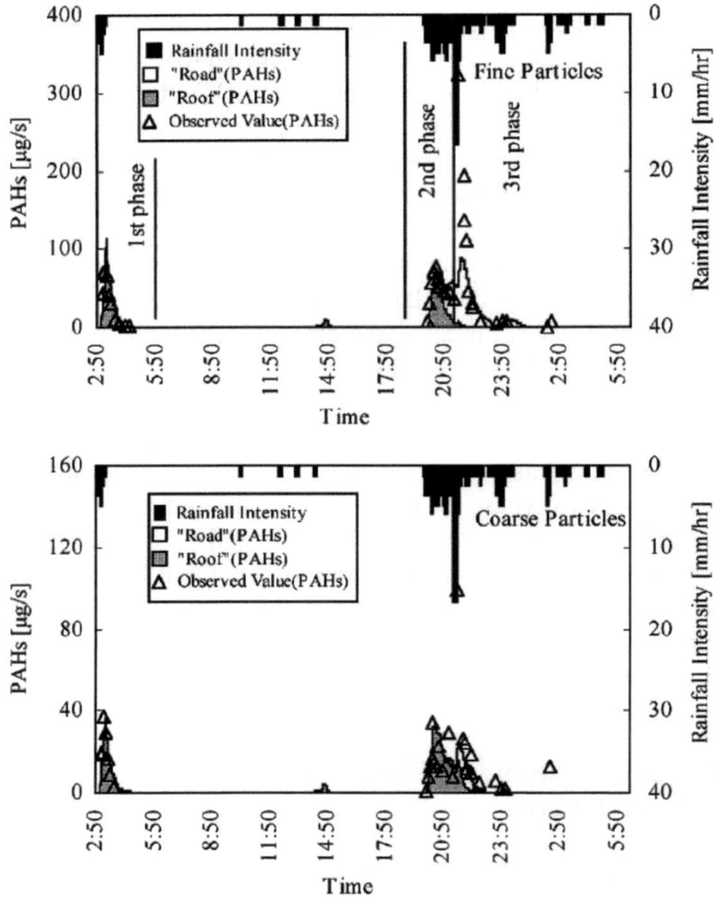

Fig. 2-3. Simulated and observed runoff of particle-associated in residential area (Murakami et al. 2004)

The bioaccessibility of hydrophobic chemicals in contaminated sediments has been tested by several researchers. An in vitro extraction using real gut fluid of benthic organisms has been proposed (Mayer et al. 1996). In a study using polychaetes (Ahrens et al. 2001), the amount desorbed in the gut fluid was almost equal to the amount assimilated into the polychaete's tissues. The best way to evaluate bioaccessibility is to employ the real gut fluid of benthic organisms, but collecting a sufficient amount of real gut fluid for the evaluation of multiple sediment samples is not realistic since only a small volume of gut fluid exists in each individual organism, and dissection of the animals is laborious. Therefore, the most desirable solution is the use of a synthetic solution as a model gut fluid. The function of the gut fluid is complex, and therefore extremely difficult to mimic completely. According to previous studies on the dissolution mechanism and its application to environmental samples (Voparil and Mayer 2000; Ahrens et al. 2001; Nakajima et al. 2005), a surfactant, namely sodium dodecyl sulfate (SDS), solution is an ideal candidate for the solvent used in in vitro bioaccessibility tests. Attempts to identify the chemical structures of real gut fluid surfactant have been progressing (Smoot et al. 2003) and a new cocktail of sodium taurocholate and bovine serum albumin was proposed to mimic polychaete gut fluid (Voparil and Mayer 2004).

We quantitatively compared the amount of PAHs extracted, using an organic solvent and by an SDS solution (as a hypothetical digestive gut fluid of polychaetes), from sediments collected from an urban stream system in Denmark receiving urban runoff (Nakajima et al. 2005). The bioaccessibility of the 12 total PAHs in the sediments was in the range of 14–38%. Lower molecular PAHs showed higher bioaccessibility compared to the higher molecular ones. The extracts from the sediments, via an organic solvent and the synthetic gut fluid, were also applied to bacterial acute toxicity tests (*Vibrio fischeri*) and algal growth inhibition tests (*Pseudokirchneriella subcapitata*). The SDS extracts showed either similar or higher degrees of toxicity than the organic solvent extracts, in spite of lower PAHs content.

The bioaccessibility of PAHs in road dust was also reported (Nakajima et al. 2006) (Fig. 2-4). They also reported PAH concentration in benthic organisms (polychaetes) and in the surrounding sediment. It was revealed that PAH composition in polychaetes did not correspond with that of the surrounding sediment. The SDS extract, or hypothetically bioaccessible fraction, of road dust selectively contained lower molecular weight PAHs, such as pyrene and fluoranthene, which were detected in the bodies of polychaetes.

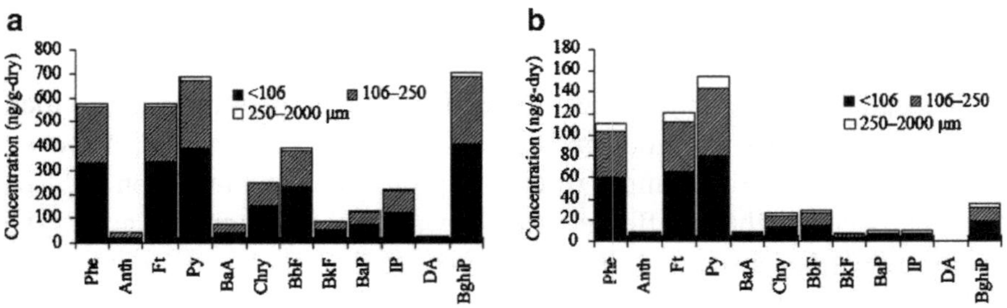

Fig. 2-4. Polycyclic aromatic hydrocarbons in road dust collected in Tokyo, Japan. (**a**) Dichloromethane-extracted fraction (total PAHs in the dust), (**b**) 1% SDS-extracted fraction (bioaccessible portion) (Nakajima et al. 2006)

However, the link between the source and the bioaccessibility is still not clear and further investigation is necessary. Although the PAH bioaccessibility in road dust was low (Nakajima et al. 2006), the alteration in the environment has yet to be evaluated.

2.2 Long-term Water Quality Monitoring in Tokyo Bay Focusing on Combined Sewer Overflow Phenomena

2.2.1 Introduction

Separate sewer systems have been installed in many cities in Japan since 1970 to improve water pollution control in public water bodies. However, as of March 2004, combined sewer systems are in place for historical reasons in 191 out of 2,246 local authorities in Japan (8.5%), including large cities such as Tokyo or Osaka. Combined sewer systems receive rainwater as well as municipal wastewater during wet weather conditions. When the flow rate exceeds the capacity of the sewer or wastewater treatment plant, the excess water overflows into public water bodies. This phenomenon is called combined sewer overflow (CSO) and has been recognized as a serious source of environmental water pollution. However, very limited field surveys have been conducted to evaluate the magnitude of the problem and the duration of its impact on receiving waters in Japan, where rather conventional and less informative water quality parameters have been monitored to assess risk of infection. Therefore, additional monitoring data and high-quality information are required to estimate the impact of CSO events on human health.

Studies in the United States investigated the relationship between waterborne infectious diseases and rain events using public records from 1971 to 1994 (Rose et al. 2000, 2001). They concluded that 20–40% of disease instances were due to contamination during heavy rain events, suggesting that public health could be improved by upgrading rainwater management systems in urban areas.

The traditional bacterial indicators for fecal contamination, such as total and fecal coliforms, are good indicators of fecal contamination; however, they are less reliable as an index of viral or protozoan contamination because these pathogens behave differently in the water environment. To assess the risk of infection caused by CSO events, both traditional bacterial indicators and viruses should be monitored in the receiving water bodies.

In this section, two monitoring data sets are introduced as advanced trials to investigate the fate of pathogens and indicator bacteria in Tokyo Bay. One is spatial and temporal monitoring, in which samples were taken in the coastal area in Tokyo. The other is long-term sampling from one site to investigate the profile of bacterial indicators and viruses.

2.2.2 Spatial and Temporal Impact of Combined Sewer Overflow

The magnitude and duration of the impact of CSO events were evaluated spatially and temporally by monitoring contaminant levels in receiving waters after rain events. In particular, the fate of enteric viruses that cause gastroenteritis (noroviruses G1 and G2 and enteroviruses) was investigated. The fates of the viruses as well as conventional indicators were evaluated by serially monitoring the receiving water body after CSO events.

Water samples were collected at 19 points in Tokyo Bay at 1 day, 2 days and 4 days after an overflow event on 1 Oct. 2002 (Katayama et al. 2004). Another set of samples was collected on 3 Aug. 2002 at 1 day after an overflow event. A set of dry weather samples was collected on 14 Dec. 2002 for reference. The sampling points are shown in Fig. 2-5. The features of the CSO events prior to the sampling are summarized in Table 2-2. The antecedent dry weather period of the CSO event on 1 Oct. was not long enough to ensure impact from the non-point sources included in the discharged water.

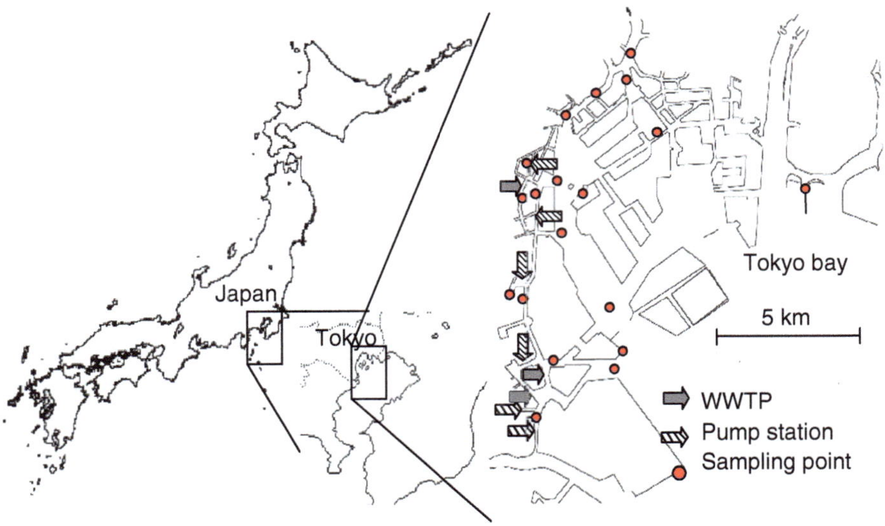

Fig. 2-5. Sampling points in Tokyo bay

Table 2-2. Features of CSO events prior to the sampling

Date of CSO event	Antecedent dry weather period (hours)	Rainfall period	Total precipitation (mm)
2 Aug. 2002	187	4 h from 16:00 on 2 Aug.	10
30 Sep. – 1 Oct. 2002	46	39 h from 9:00 on 30 Sep.	66.5

All samples were assayed for total coliforms by the double agar layer method using deoxycholic acid agar. Fecal coliforms were tested by the same method except for incubating at 44.5±0.2°C. Viruses were concentrated from 2 L of each sample and identified following the procedure described in Katayama et al. (2002). Consequently, volume equivalent to 300 mL of an initial sample was tested in a direct reverse transcriptase-polymerase chain reaction (RT-PCR) to detect norovirus types G1 and G2 and enteroviruses. Cell culture–RT-PCR was also performed to detect infectious enteroviruses using BGM cell.

Table 2-3 presents data on the water quality parameters of temperature, *p*H and salinity. Seasonal effects are reflected in temperature, and the effects of precipitation can be observed in the lower salinity compared to the dry day.

Occurrences of enteric viruses and the levels of total and fecal coliforms in Tokyo Bay after CSO events are shown in Table 2-4. The samples on 3 Aug. included more positive results of enterovirus than those on 2 Oct. Either of the two enteric viruses was detected in coastal water samples from 6 out of 19 sites, 10 out of 12 sites (data not shown) and 8 out of 14 sites on 2 Oct., 3 Oct. and 5 Oct., respectively, showing almost no decrease over this 4-day period. This tendency was also observed in the culturable enteroviruses, which was detected in 2 out of 19 sampling points, 8 out of 12 points and 7 out of 14 points on 2 Oct., 3 Oct. and 5 Oct., respectively. Positive ratio of the noroviruses were not decreased from 2 Oct. to 5 Oct. In contrast, the levels of total and fecal coliforms decreased by one order of magnitude from 2 Oct. to 3 Oct.

Figure 2-6 shows the relationship between total and fecal coliforms. Black spots indicate the presence of enteroviruses (a) and noroviruses (b) in the corresponding samples. A good correlation between total and fecal coliforms was observed, while both enteroviruses and noroviruses were detected even when levels of the two bacterial indicators were low. If enteric viruses occurred proportionally with bacterial counts, one would expect to see no black spots in areas of the graph where coliform counts are very low.

Table 2-3. Range of water quality parameters after CSO events and during dry weather

Sampling date (day after CSO)	Temperature (°C)	*p*H	Salinity (‰)
3 Aug. (day 1)	27.9–31.0	6.93–7.36	3.7–18.7
2 Oct. (day 1)	19.4–23.7	6.69–8.28	0.3–15.5
3 Oct. (day 2)	20.9–24.5	7.24–7.79	1.8–13.1
5 Oct. (day 4)	21.8–25.7	7.06–7.76	6.0–19.7
4 Dec. (dry day)	10.5–15.6	7.13–7.79	11.1–27.9

Table 2-4. Occurrences of enteric viruses in Tokyo Bay after CSO events

Sampling date	NV[a]	EV[b]	Culturable EV	TC[c]	FC[c]
3 Aug. (day 1)	6/19	9/19	N. D.[d]	$10^{3.54\pm0.69}$	N. D.[d]
2 Oct. (day 1)	4/19	6/19	2/19	$10^{3.42\pm0.61}$	$10^{2.84\pm0.55}$
3 Oct. (day 2)	3/12	8/12	8/12	$10^{2.49\pm0.46}$	$10^{1.81\pm0.42}$
5 Oct. (day 4)	3/14	8/14	7/14	$10^{2.54\pm0.62}$	$10^{1.75\pm0.57}$

[a] No. of samples positive for either G1 or G2 / No. of samples tested
[b] No. of samples positive for EV by either direct RT-PCR or cell culture–RT-PCR / No. of samples tested
[c] Average and standard deviation of log normal distribution
[d] No data

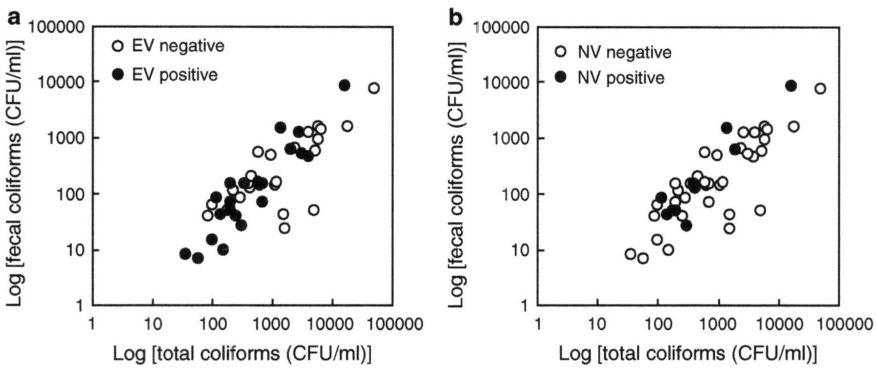

Fig. 2-6. Relationships among total coliforms, fecal coliforms and enterovirus (**a**) and noroviruses (**b**)

These data show that it is not reliable to assess the risk of infection of enteric viruses from the level of the two bacterial indicators in Tokyo Bay.

The number of samples testing positive for enteric viruses in Tokyo Bay did not increase or decrease according to their positive ratio at day 1, 2 and

4 after the CSO event. On the other hand, the number of total and fecal coliforms increased rapidly 1 day after the CSO event and had decreased drastically by days 2 and 4. The origin, behaviour and fate of viruses in the environment may be different from those of bacterial indicators. The importance of measuring enteric viruses together with the bacterial indicators was confirmed by this study.

2.2.3 Long-term Monitoring of Impact of Combined Sewer Overflow

The correlation of bacterial indicators and viruses could not be evaluated in the previous section because the quantitative concentration of viruses was not determined. To compare the behaviour of viruses and indicators, the duration and frequency of sampling were prioritized using only one sampling site and determining limited parameters. A 2-month survey was conducted to evaluate the effects of rainfall on the fate of human adenoviruses, total coliforms and *Escherichia coli* in coastal water in the Odaiba area in Tokyo Bay (Haramoto et al. 2006).

The Odaiba area is suspected to be contaminated with the effluents from several domestic wastewater treatment plants. The Odaiba Marine Park is located near the sampling site, and more than one million people visit the park for recreational purposes annually. Although playing on the beaches is allowed, swimming in the sea is prohibited.

2.2.3.1 Materials and Methods

Samples were collected on 47 of 73 days from 4 Aug. to 15 Oct. 2004. During the survey period, a total of 774 mm of rainfall was observed, including some heavy rainfall events caused by typhoons. Samples were usually collected in the morning, delivered to the laboratory within a few hours on ice and analysed for human adenoviruses, total coliforms and *E. coli*.

The acid rinse method (Katayama et al. 2002) was used for concentrating the virus from 1,000 ml of coastal water samples as in the previously described study. Viral DNA was extracted using commercially available methods, followed by PCR to determine the concentration of viral genomes. At the same time, a decimal dilution series of DNA from human adenovirus serotype 40 were used to create a calibration curve.

Total coliforms and *E. coli* in 10 mL of coastal water were determined by an m-Coliblue broth membrane filtration procedure (Millipore, Tokyo).

2.2.3.2 Results

Total coliforms and *E. coli* were detected in all 47 tested samples with geometric mean concentrations of 68 CFU/mL (range: 1.8–3,700 CFU/mL) and 4.4 CFU/mL (range: 0.15–280 CFU/mL), respectively. On the other hand, human adenoviruses were detected in 38 (81%) of 47 samples at a maximum concentration of 5.5 PDU/mL.

The profiles of the concentrations of human adenoviruses, total coliforms and *E. coli* in coastal water are shown in Fig. 2-7 (from 23 Aug. to 10 Sep.). The concentrations of these microorganisms increased after rainfall events. For instance, following a heavy rainfall from 4 to 5 Sept. (84.5 mm), the concentration of human adenoviruses, total coliforms and *E. coli* increased from 0.14 to 5.5 PDU/mL, from 13 to 240 CFU/mL and from 2.0 to 55 CFU/mL, respectively. These increased concentrations decreased gradually to the level before the rainfall event within a few days to 0.24 PDU/mL, 21 CFU/mL and 1.9 CFU/mL, respectively, on 8 Sept.

Tidal effects on the occurrence of the microorganism are shown in Fig. 2-8. All the samples were divided into two groups depending on the tidal movement at the sampling time: an increasing-tide group or a decreasing-tide group. For all tested microorganisms, there was no significant difference in

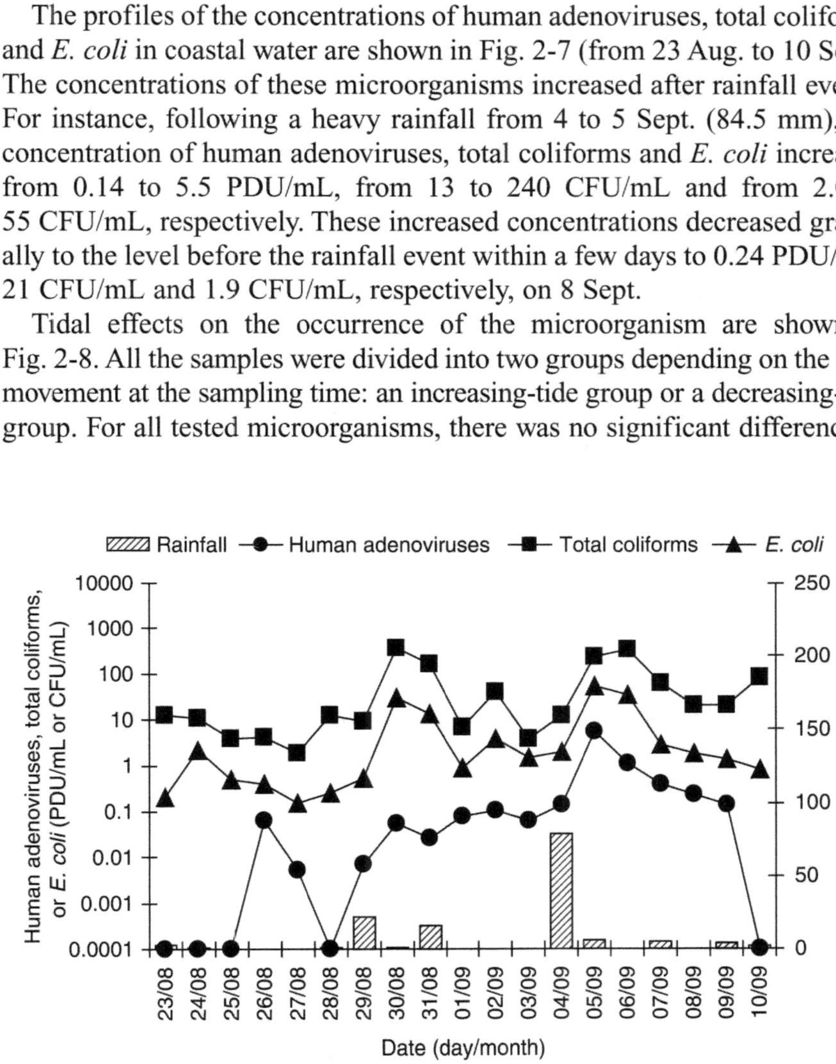

Fig. 2-7. Occurrence of human adenoviruses, total coliforms and *E. coli* in coastal water (from 23 Aug. to 10 Sept. 2004) (Haramoto et al. 2006)

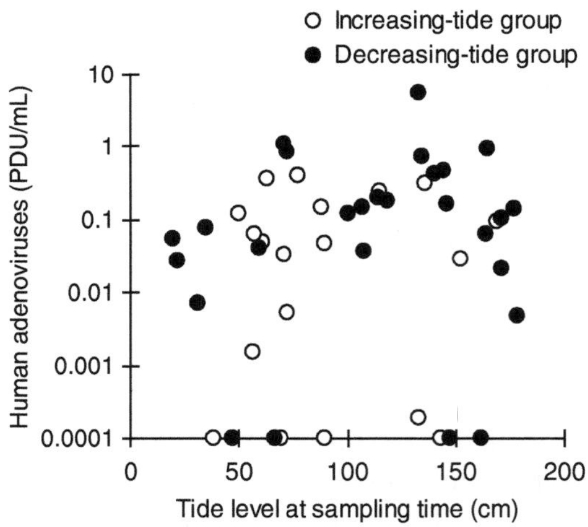

Fig. 2-8. Relationship between tide level and concentration of human adenoviruses (Haramoto et al. 2006)

the concentration between the increasing-tide group and the decreasing-tide group (*t*-test, $P>0.05$).

The relationship between the concentration of human adenoviruses and that of total coliforms or *E. coli* was determined. As shown in Fig. 2-9, a moderate positive correlation ($r=0.536$) was observed between the logarithms of the concentration of human adenoviruses and that of *E. coli* among the adenovirus-positive samples.

2.2.3.3 Discussion

Human adenoviruses were chosen as a target virus, and total coliforms and *E. coli* were used as indicator bacteria to compare the fates of viruses and bacteria in CSO-contaminated coastal water. After a rainfall event, the concentration of the tested microorganisms in coastal water usually increased by 10–100 times, followed by a gradual decrease to the level before the rainfall event within a few days.

Total coliforms and *E. coli* are present in the feces of both humans and animals, while human adenoviruses are present only in human feces. The results of this study indicate that untreated sewage, or CSO, could be a major source of these microorganisms in Tokyo Bay. Interestingly, this study indicated that the high contamination of human adenoviruses following heavy rainfall persisted for at least a few days after the event. Accordingly, recreational activities in the contaminated area after rain will pose a higher risk for infection by human adenoviruses.

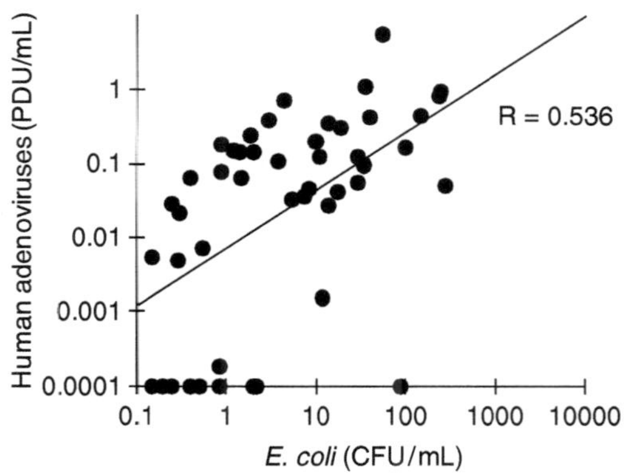

Fig. 2-9. Relationships between concentrations of human adenoviruses and *E. coli* in coastal water samples from the same location on several days (Haramoto et al. 2006)

Coastal water samples in the decreasing-tide group were expected to contain a higher proportion of effluents from wastewater treatment plants than those in the increasing-tide group. In this study, no significant difference was observed in the concentration of microorganisms between the increasing- and decreasing-tide groups (Fig. 2-8), suggesting that the impact of rainfall on the fate of microorganisms is stronger than that of tidal movement.

A moderate positive correlation was found between the concentration of human adenoviruses and that of *E. coli* (Fig. 2-9), which suggests that *E. coli* could be used as an indicator of human adenovirus contamination in coastal water susceptible to CSO.

2.2.4 Summary

The concentration of tested microorganisms increased after a rainfall event and then gradually decreased to the level before the rainfall event within a few days. There was no significant difference in the concentration of the tested microorganisms between the increasing- and decreasing-tide groups, suggesting that the impact of rainfall on the fate of microorganisms is stronger than that of tidal movement.

The behaviours of a pathogenic virus as well as bacterial indicators were gradually elucidated by the study of intensive sampling. Techniques of molecular biology and virus concentration enabled us to study the fate of human pathogenic viruses in a CSO-impacted water body.

References

Ahrens MJ, Hertz J, Lamoureux EM, Lopez GR, McElroy AE, Brownawell BJ (2001) The role of digestive surfactants in determining bioavailability of sediment-bound hydrophobic organic contaminants to 2 deposit-feeding polychaetes. Mar Ecol Prog Ser 212:145–157

Andral MC, Roger S, Montrejaud-Vignoles M, Herremans L (1999) Particle size distribution and hydrodynamic characteristics of solid matter carried by runoff from motorways. Water Environ Res 71:398–407

Baumann PC (1998) Epizootics of cancer in fish associated with genotoxins in sediment and water. Mutation Res/Rev Mutation Res 411:227–233

Boxall ABA, Maltby L (1997) The effects of motorway runoff on freshwater ecosystems: 3. Toxicant confirmation. Arch Environ Contam Toxicol 33:9–16

Brenner RC, Magar VS, Ickes JA, Abbott JE, Stout SA, Crecelius EA, Bingler LS (2002) Characterization and fate of PAH-contaminated sediments at the Wyckoff/Eagle Harbor superfund site. Environ Sci Technol 36:2605–2613

Brown RC, Pierce RH, Rice SA (1985) Hydrocarbon contamination in sediments from urban stormwater runoff. Mar Poll Bull 16:236–240

Chebbo G, Gromaire MC, Ahyerre M, Garnaud S (2001) Production and transport of urban wet weather pollution in combined sewer system: the "Marais" experimental urban catchment in Paris. Urban Water 3:3–15

Eriksson E, Baun A, Mikkelsen PS, Ledin A (2005) Chemical hazard identification and assessment tool for evaluation of stormwater priority pollutants. Water Sci Technol 51(2):47–55

Förster J (1999) Variability of roof runoff quality. Water Sci Technol 39(5):137–144

Furumai H, Hijioka Y, Nakajima F (2001) Modelling and field survey on washoff behaviour of suspended particles from roofs and roads. In: Brashear RW, Maksimovic C (eds) Urban drainage modelling – Proceedings of the specialty symposium of the world water and environmental resources congress. American Society of Civil Engineers, pp 225–237

Furumai H, Balmer H, Boller M (2002) Dynamic behavior of suspended pollutants and particle size distribution in highway runoff. Water Sci Technol 46(11–12): 413–418

Ghosh U, Gillette JS, Luthy RG, Zare RN (2000) Microscale location, characterization, and association of polycyclic aromatic hydrocarbons on harbor sediment particles. Environ Sci Technol 34:1729–1736

Ghosh U, Zimmerman JR, Luthy RG (2003) PCB and PAH speciation among particle types in contaminated harbor sediments and effects on PAH bioavailability. Environ Sci Technol 37:2209–2217

Hagris WJ Jr, Roberts MH Jr, Zwerner DE (1984) Effects of contaminated sediments and sediment-exposed effluent water on an estuarine fish: acute toxicity. Marine Environ Res 14:337–354

Haramoto E, Katayama H, Oguma K, Koibuchi Y, Furumai H, Ohgaki S (2006) Effects of rainfall on the occurrence of human adenoviruses, total coliforms, and Escherichia coli in seawater. Water Sci Technol 54(3):225–230

Heaney JP, Pitt R, Field R (1999) Innovative urban wet weather flow management systems. EPA/600/R-99/029. US Environmental Protection Agency, Office of Research and Development, National Risk Management Research Laboratory, Cincinnaty, OH

Hijioka Y, Nakajima F, Furumai H (2001) Modified models of wash-off from roofs and roads for non-point pollution analysis during first flush phenomena. In: Brashear RW, Maksimovic C (eds) Urban drainage modelling – Proceedings of the specialty symposium of the world water and environmental resources congress. American Society of Civil Engineers, pp275–286

Hoffman EJ, Mills GL, Latimer JS, Quinn JG (1984) Urban runoff as a source of polycyclic aromatic hydrocarbons to coastal waters. Environ Sci Technol 18:580–587

Katayama H, Shimasaki A, Ohgaki S (2002) Development of a virus concentration method and its application to detection of enterovirus and Norwalk virus from coastal seawater. Appl Environ Microbiol 68:1033–1039

Katayama H, Okuma K, Furumai H, Ohgaki S (2004) Series of Surveys for Enteric Viruses and Indicator Organisms in Tokyo Bay after an Event of Combined Sewer Overflow. Water Science & Technology 50(1):259–262

Kelsey J, Alexander M (1997) Declining bioavailability and inappropriate estimation of risk of persistent compounds. Environ Toxicol Chem 16(3):582–585

Krein A, Schorer M (2000) Road runoff pollution by polycyclic aromatic hydrocarbons and its contribution to river sediments. Water Res 34:4110–4115

Maltby L, Forrow DM, Boxall ABA, Calow P, Betton CI (1995a) The effects of motorway runoff on freshwater ecosystems: 1. Field study. Environ Toxicol Chem 14(6):1079–1092

Maltby L, Boxall ABA, Forrow DM, Calow P, Betton CI (1995b) The effects of motorway runoff on freshwater ecosystems: 2. Identifying major toxicants. Environ Toxicol Chem 14(6):1093–1101

Mayer LM, Chen Z, Findlay RH, Fang J, Sampson S, Self RF, Jumars PA, Quetel C, Donard OFX (1996) Bioavailability of sedimentary contaminants subject to deposit-feeder digestion. Environ Sci Technol 30(8):2641–2645

Murakami M, Nakajima F, Furumai H (2003) Distinction of size-fractionated road and roof dust based on PAHs contents and profiles. J Japan Society on Water Environ 26(12):837–842 (Main text in Japanese; with the English abstract and the figures and tables in English)

Murakami M, Nakajima F, Furumai H (2004) Modelling of runoff behaviour of particle-bound polycyclic aromatic hydrocarbons (PAHs) from roads and roofs. Water Res 38(20):4475–4483

Murakami M, Nakajima F, Furumai H (2005) Size- and density-distributions and sources of polycyclic aromatic hydrocarbons in urban road dust. Chemosphere 61:783–791

Nakajima F, Baun A, Ledin A, Mikkelsen PS (2005) A novel method for evaluating bioavailability of polycyclic aromatic hydrocarbons in sediments of an urban stream. Water Sci Technol 51(3–4):275–281

Nakajima F, Saito K, Isozaki Y, Furumai H, Christensen AM, Baun A, Ledin A, Mikkelsen PS (2006) Transfer of hydrophobic contaminants in urban runoff particles to benthic organisms estimated by an in vitro bioaccessibility test. Water Sci Technol 54(6–7):323–330

Pengchai P, Furumai H, Nakajima F (2004) Source apportionment of polycyclic aromatic hydrocarbons in road dust in Tokyo. Polycycl Aromat Compd 24(4–5): 713–789

Pengchai P, Nakajima F, Furumai H (2005) Estimation of origins of polycyclic aromatic hydrocarbons in size-fractionated road dust in Tokyo with multivariate analysis. Water Sci Technol 51(3–4):169–175

Phillips DH (1983) Fifty years of benzo(a)pyrene. Nature 303:468–472

Rockne KJ, Shor LM, Young LY, Taghon GL, Kosson DS (2002) Distributed sequestration and release of PAHs in weathered sediment: The role of sediment structure and organic carbon properties. Environ Sci Technol 36:2636–2644

Roger S, Montrejaud-Vignoles M, Andral MC, Herremans L, Fortune JP (1998) Mineral, physical and chemical analysis of the solid matter carried by motorway runoff water. Water Res 32:1119–1125

Rose JB, Daeschner S, Easterling DR, Curriero FC, Lele S, Patz JA (2000) Climate and waterborne disease outbreaks. J Am Water Works Assoc 92(9):77–87

Rose JB, Epstein PR, Lipp EK, Sherman BH, Bernard SM, Patz JA (2001) Climate variability and change in the United States: Potential impacts on water- and foodborne diseases caused by microbiologic agents. Environ Health Perspect 109(5):211–221

Sansalone JJ, Buchberger SG (1997) Characterization of solid metal element distributions in urban highway stormwater. Water Sci Technol 36(8–9):155–160

Sartor JD, Boyd GB (1972) Water pollution aspects of street surface contaminants. EPA/R2/72/081. US Environmental Protection Agency, Office of Research and Monitoring, Washington, DC

Smoot JC, Mayer LM, Bock MJ, Wood PC, Findlay RH (2003) Structures and concentrations of surfactants in gut fluid of the marine polychaete *Arenicola marina*. Mar Ecol Prog Ser 258:161–169

Takada H, Onda T, Ogura N (1990) Determination of polycyclic aromatic hydrocarbons in urban street dusts and their source materials by capillary gas chromatography. Environ Sci Technol 24:1179–1186

Tomonvic A, Makishimovic C (1996) Improved modeling of suspended solids discharge from asphalt surface during storm event. Water Sci Technol 33(4–5):363–369

Uchimura K, Nakamura E, Fujita S (1997) Characteristics of stormwater runoff and its control in Japan. Water Sci Technol 36(8–9):141–147

Voparil IM, Mayer LM (2000) Dissolution of sedimentary polycyclic aromatic hydrocarbons into the lugworm's (*Arenicola marina*) digestive fluids. Environ Sci Technol 34(7):1221–1228

Voparil IM, Mayer LM (2004) Commercially available chemicals that mimic a deposit feeder's (*Arenicola marina*) digestive solubilization of lipids. Environ Sci Technol 38(16):4334–4339

Zakaria MP, Takada H, Tsutsumi S, Ohno K, Yamada J, Kouno E, Kumata H (2002) Distribution of polycyclic aromatic hydrocarbons (PAHs) in rivers and estuaries in Malaysia: a widespread input of petrogenic PAHs. Environ Sci Technol 36:1907–1918

3. Numerical Simulation of Urban Coastal Zones

Yukio Koibuchi and Shinji Sato

3.1 Numerical Modeling in Urban Coastal Zones

Numerical modeling is an essential technique for the understanding and management of water quality in urban coastal zones. This may be because urban coastal zones are characterized by an extremely wide variety of environments with complicated geographical features, such as urban areas that intersect with outer oceans, and hence are affected by both. The phenomena in this area are not only physical, but also biological or chemical, and they interfere with each other. Therefore, the ecosystem and the water quality of urban coastal zones are highly complicated.

The quality of water has long been deteriorating at many of the world's urban coastal zones (Walker 1990). Damage caused by events such as red tides, harmful algal bloom, and decreased amounts of dissolved oxygen in bottom water occur frequently. Such phenomena induce the degradation of aquatic ecosystems and the loss of aquatic resources such as sea grass beds, as well as fish and shellfish. These phenomena – caused by an excessive inflow of nutrients that has been accelerated by urban populations, resulting in the increase of concentrations called eutrophication (Caperon et al. 1971) – have been causing many problems for fisheries and recreation areas in urban coastal zones. However, these events are also influenced by weather conditions, ocean currents, household and industrial wastes, nutrients coming from agricultural lands, sewage treatment plants, acid rain and other such phenomena (Ærtebjerg et al. 2003). Furthermore, the interactions between all of these factors are quite complicated. Actually, distinguishing between natural and anthropogenic effects is not easy (Jøgensen and Richardson 1996). This makes the improvement of water quality and the management of resources in this area even more difficult.

H. Furumai et al. (eds.), *Advanced Monitoring and Numerical Analysis of Coastal Water and Urban Air Environment*,

© 2010 to the complete printed work by Springer, except as noted. Individual authors or their assignees retain rights to their respective contributions; reproduced by permission.

Numerical modeling that deals quantitatively with these complex systems has been in development since the 1960s, much like digital computers. The use of modeling also has important practical applications, such as the prediction and management of water quality and ecosystems in urban coastal zone.

This chapter describes how water quality and pathogens are modeled in urban coastal areas. Sect. 3.1.2 explains the basic structure of these models. Physical and ecosystem modeling techniques for enclosed bays are described in Sect. 3.1.3. Sect. 3.1.4 shows the application of ecosystem modeling as a tool for integrated management.

Next, Sect. 3.2 describes pathogen modeling in urban coastal areas.

3.1.1 Introduction to Physical Numerical Modeling in Urban Coastal Areas

Water quality is directly and indirectly influenced by a variety of flows. For example, since phytoplankton drifts passively according to water flows, it is strongly influenced by the distribution of the flow of the bay (Lucas et al. 1999). Bay water motions also exert a strong influence on other water quality parameters. These kinds of influences are direct and obvious. The degree of influence naturally increases with the strength of the currents.

In contrast, even if the flow is very small, it has some spatial patterns in common with outer bay directions. In this case, a nutrient load that is discharged at the head of the bay is transported to distant locations and, probably, the water exchange rate of the bay also increases. The nutrient loading that is permissible – i.e. which the bay can accept – will then increase due to the increase in the water exchange rate. For water retention rates, the strength of currents is not as important as spatial patterns. For example, tidal currents are dominant in urban coastal areas. However they are oscillatory. Water particles in tidal currents move to the head of the bay during flood tides, but move back to the mouth of the bay during ebb tides. As a result, tidal currents do not substantially transport water particles. In contrast, density currents are clearly weaker than tidal currents, but they flow in one direction continuously and transport water particles more efficiently. As a result, they substantially effect water retention times and ecosystem characteristics.

A water quality model for bays must therefore consist of a three-dimensional circulation model and an ecosystem model that describes pelagic and benthic aspect of nutrients cycling. This section focuses on the physical modeling and the next section deals with water quality modeling. The final section discusses the application of these models to Tokyo Bay.

3.1.2 Three-Dimensional Hydrodynamic Model

Many three-dimensional hydrodynamic models have been developed in the last decades, including POM (Princeton Ocean Model; Blumberg and Mellor 1987), CH3D (Curvilinear Hydrodynamics in 3 Dimensions; Johnson et al. 1993) and ROMS (Regional Ocean Modeling System; MacCready et al. 2002; Li et al. 2005). These models solve the Navier-Stokes equation with the forcing (the wind stress, Coriolis force and buoyancy force) under adequate approximations that are called the hydrostatic and Boussinesq approximations.

Hydrostatic approximation assumes that there is a perfect balance between pressure gradients and gravity: in other words, no acceleration occurs in a vertical direction. This is justified because the aspect ratio of urban coastal areas is extremely small, and hence the vertical motions are considered to be small and also to be further inhibited by gravitational forces under stable density stratification. This means that vertical acceleration is negligible and the fluid behaves as though it were under static equilibrium as far as vertical motion is concerned (Prudman 1953).

Density variations in urban coastal areas are also small (less than 3% or so), and so density can be considered to be constant, except when body forces resulting from the motion of a density stratified fluid in a gravitational field are concerned. This approximation is called the Boussinesq approximation: in other words, changes in the mass or inertia of a fluid body due to the changes in its density are negligible, while the same changes in density are consequential when the gravitational field is present (Kantha and Clayson 2000). Therefore, following Boussinesq (1903), this approximation justifies replacing ρ by a constant reference density ρ_0 everywhere except in terms involving gravitational acceleration constant g. Under such approximations, the governing equations are transformed as follows:

$$\frac{\partial u}{\partial t}+u\frac{\partial u}{\partial x}+v\frac{\partial u}{\partial y}+w\frac{\partial u}{\partial z}$$
$$=fv-g\frac{\partial \eta}{\partial x}-\frac{g}{\rho_0}\frac{\partial}{\partial x}\int_z^{\eta}\rho' g dz+\frac{\partial}{\partial x}(A_x\frac{\partial u}{\partial x})+\frac{\partial}{\partial y}(A_y\frac{\partial u}{\partial y})+\frac{\partial}{\partial z}(A_z\frac{\partial u}{\partial z}) \quad (3.1)$$

$$\frac{\partial v}{\partial t}+u\frac{\partial v}{\partial x}+v\frac{\partial v}{\partial y}+w\frac{\partial v}{\partial z}$$
$$= fu-g\frac{\partial \eta}{\partial y}-\frac{g}{\rho_0}\frac{\partial}{\partial y}\int_z^{\eta}\rho' g dz+\frac{\partial}{\partial x}(A_x\frac{\partial v}{\partial x})+\frac{\partial}{\partial y}(A_y\frac{\partial v}{\partial y})+\frac{\partial}{\partial z}(A_z\frac{\partial v}{\partial z}) \tag{3.2}$$

$$\frac{\partial u}{\partial x}+\frac{\partial v}{\partial y}+\frac{\partial w}{\partial z}=0 \tag{3.3}$$

Here, t stands for time. u, v, and w are the velocity components in x, y, and z directions. The symbol ρ' is the reference density and it is defined as $\rho = \rho_0 + \rho'$. The symbols A_x, A_y and A_z are the eddy viscosities in x, y, and z directions. The symbol g is the acceleration due to gravity.

3.1.2.1 Grid Systems of Urban Costal Model

All currents including tides, wind-driven currents, and density currents in urban coastal zones are strongly influenced by geometry and bathymetry, whereas these areas are rarely regular in shape. In particular, a coastline near an urban coastal area is more complex, due to reclamations and the constructions of harbors, than a natural one. In addition, the uniformity of the bathymetry further lessened as a result of dredging for vessel transport. A computational grid is required to accurately represent such complex geometry and bathymetry. For this reason, the selection of which grid system to use has varied along with the progress of modeling, although the governing equations are not basically different.

For vertical coordinate systems (shown in Fig. 3-1), Cartesian (z-coordinate vertical grid) and sigma-coordinate grids have been widely used. A Cartesian grid is easy to understand, and shows the correspondence between program codes and governing equations. It is sometimes more accurate than sigma-coordinate grids, especially if the bathymetry of the bay is simple and mild. The sigma-coordinate system tends to have an error featuring the presence of steep-bottom topography. However, unless an excessively large number of vertical levels are employed, the Cartesian grid fails to represent the bottom topography with satisfied accuracy.

The sigma-coordinate system is convenient in the sense that it can essentially introduce a "flattening out" mechanism for variable bottoms at $z = -h(x, y)$. The flow near the seabed is also calculated well. Moreover, the sigma-coordinate system is easy to program since the number of vertical grids can be the same, and the setting of boundary conditions will be simple. The sigma-coordinate system has long been widely used in both meteorology and oceanography (Phillips 1957; Freeman et al. 1972).

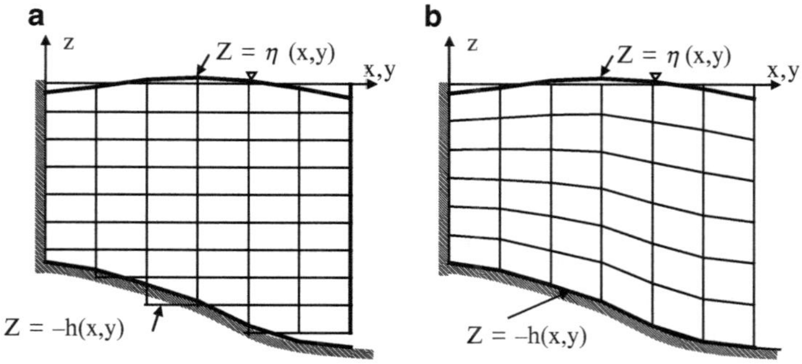

Fig. 3-1. Examples of two vertical grid systems: (**a**) a Cartesian or *z*-coordinate vertical grid and (**b**) a sigma-coordinate vertical grid system

After the incorporation of approximations, the governing equations in the sigma coordinate system are as follows:

$$\frac{\partial(Hu)}{\partial t}+\frac{\partial(Huu)}{\partial x}+\frac{\partial(Hvu)}{\partial y}+\frac{\partial(H\dot{\sigma}u)}{\partial\sigma} \tag{3.4}$$

$$= Hfv-\frac{H}{\rho}\frac{\partial p}{\partial x}+\frac{1}{H}\frac{\partial}{\partial\sigma}(A_v\frac{\partial u}{\partial\sigma})+HA_h\left(\frac{\partial^2 u}{\partial x^2}+\frac{\partial^2 u}{\partial y^2}\right)$$

$$\frac{\partial(Hv)}{\partial t}+\frac{\partial(Huv)}{\partial x}+\frac{\partial(Hvv)}{\partial y}+\frac{\partial(H\dot{\sigma}v)}{\partial\sigma} \tag{3.5}$$

$$= Hf(-u)-\frac{H}{\rho}\frac{\partial p}{\partial y}+\frac{1}{H}\frac{\partial}{\partial\sigma}(A_v\frac{\partial v}{\partial\sigma})+HA_h\left(\frac{\partial^2 v}{\partial x^2}+\frac{\partial^2 v}{\partial y^2}\right)$$

$$-\frac{1}{\rho}\nabla p=-\frac{1}{\rho}\left[(\rho_0+\rho'\sigma)g\nabla_\sigma\eta+\rho' g(\sigma-1)\nabla_\sigma h+\nabla_\sigma\left[H\int_\sigma^1\rho' g d\sigma\right]\right] \tag{3.6}$$

$$\sigma=\frac{z+h}{h+\eta} \tag{3.7}$$

$$\dot{\sigma}=\frac{\partial\sigma}{\partial t}+u\frac{\partial\sigma}{\partial x}+v\frac{\partial\sigma}{\partial y}+w\frac{\partial\sigma}{\partial z} \tag{3.8}$$

where, t stands for time. u, v, and $\dot{\sigma}$ are the velocity components in x, y, and z directions in the $\dot{\sigma}$ coordinate system. η is change in surface elevation, h is initial water depth, and H is the total water depth ($H=h+\eta$). f is the Coriolis coefficient and P stands for pressure. ρ and ρ' are the constant reference density and the deviation from it, and $\rho = \rho_0 + \rho'$. A_H and A_V are the horizontal and vertical eddy viscosity coefficients, respectively. g is acceleration due to gravity.

Recently, the stretched grid system (S-grid system) has also been popular. This grid system is an extension of sigma-coordinate system. Generally, a sigma-coordinate grid divides the vertical coordinate into an equal number of points. The S-grid system has a higher resolution near the surface and bottom (Haidvogel et al. 2000).

Wind-stress and bottom friction are considered at surface and bottom boundary conditions, respectively. Settings in boundary conditions are easy in sigma-coordinate systems since they use the same number of vertical grids. At lateral boundaries, normal velocities are set at zero, and a free slip condition is applied to the friction terms. At open boundary, the velocity gradient is set at zero.

Horizontal computational grids are also modified to fit topography. The simplest horizontal computational grid is the rectangular grid with fixed spacing. The rectangular grid is equivalent to the Cartesian vertical grid. Recently, curvilinear coordinate systems have been widely used. These systems allow greater flexibility than rectangular grid systems. Fig. 3-2 is

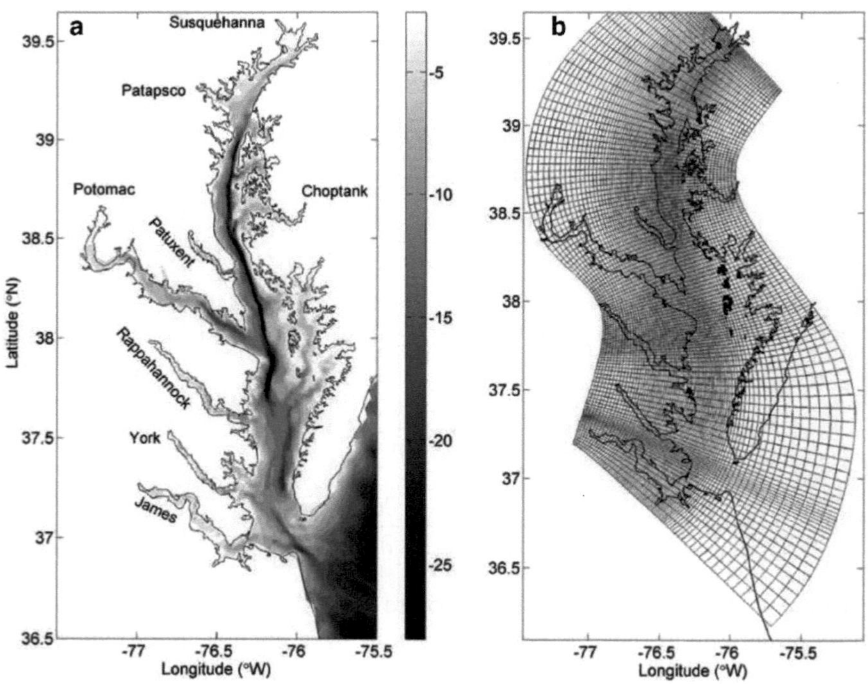

Fig. 3-2. Examples of curvilinear horizontal coordinate systems. (**a**) Bathymetry of the Chesapeake Bay and its adjacent coastal area. Depths are in meters. (**b**) A horizontal curvilinear coordinate system designed for resolving the bay's complex coastlines and deep channel (from Ming et al. 2005)

an example of a horizontal curvilinear coordinate system. The Chesapeake Bay, like other bays, has a typical complex geometry, and thus a horizontal curvilinear coordinate system is advantageous (Li et al. 2005).

This is extended to a nested grid system in which finer grids are used in regions to yield detailed information.

3.1.2.2 Density Effect Modeling for Urban Coastal Waters

In urban coastal areas, density difference plays an essential role for water quality and currents. One of the most important phenomena induced by the density effect is stratification. Once stratification occurs in a coastal zone, surface water and bottom water are isolated. This process is very important when we discuss the distributions of pollutants from the land. Stratification also enhances the increase of phytoplankton in the surface layer and oxygen depletion near the seabed. Moreover, estuarine circulation is induced by density differences in the salty sea water and the river flow. Fig. 3-3 shows a schematic diagram of estuarine circulation. River water runs through the urban area and runs off from the river mouth, spreading over the sea surface like a veil since river water has low density compared with saline sea water. To cancel the density difference between the river water and the saline water, a great deal of sea water is entrained into the river water flow. Such a mixing process continues until the river water reaches the same density as the surrounding sea water, resulting in vertical circulation in the bays that is is several to ten times greater than the river flux (Unoki 1998). Thus, estuarine circulation induces seaward currents on the surface and landward currents near the bottom. The speed of the currents is slow compared with the tidal currents, as explained previously. Since estuarine circulations are in a fixed direction, its material transport is very effective over a long time scale in spite of the small magnitude of its velocity.

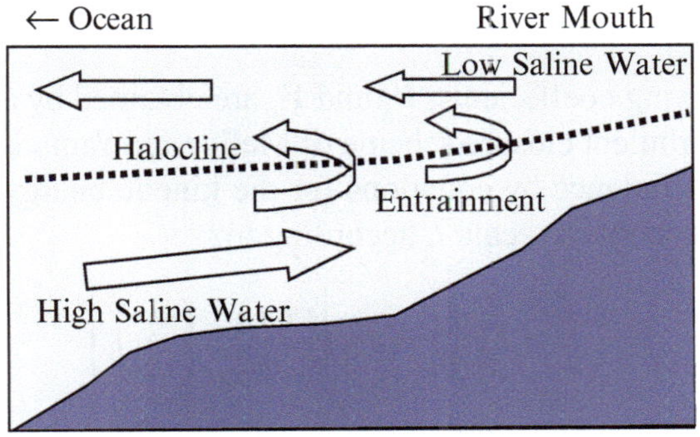

Fig. 3-3. Cross-sectional view of two-layer estuarine circulation in a bay

Estuarine circulation also plays an important role in the nutrient cycles of stratified bays. Organic matters are deposited on the seabed after phytoplankton blooms or river runoffs. They are decomposed by bacteria in the seabed. These nutrients are supplied from the seabed under anoxic condition in summer.

In order to include density effects in the numerical model, conservation equations for temperature and salinity are also included. Then, to obtain a realistic prediction for vertical stratification, a turbulent closure model is employed (Mellor and Yamada 1982). Consequently, flows driven by various mechanisms – e.g. the gravitational, wind-driven, and topographically induced flows – can be reproduced within physical numerical models.

Diffusion equations for temperature and salinity in the sigma coordinate system are as follows:

$$\frac{\partial(HT)}{\partial t}+\frac{\partial(uHT)}{\partial x}+\frac{\partial(vHT)}{\partial y}+\frac{\partial(\dot{\sigma}HT)}{\partial\sigma}= \qquad (3.9)$$

$$HK_H\left(\frac{\partial^2 T}{\partial x^2}+\frac{\partial^2 T}{\partial y^2}\right)+\frac{1}{H^2}\frac{\partial}{\partial\sigma}\left(K_V\frac{\partial HT}{\partial\sigma}\right)+\frac{1}{\rho C_P}\frac{dQ(\sigma)}{d\sigma}$$

$$\frac{\partial(HS)}{\partial t}+\frac{\partial(uHS)}{\partial x}+\frac{\partial(vHS)}{\partial y}+\frac{\partial(\dot{\sigma}HS)}{\partial\sigma}= \qquad (3.10)$$

$$HK_H\left(\frac{\partial^2 S}{\partial x^2}+\frac{\partial^2 S}{\partial y^2}\right)+\frac{1}{H^2}\frac{\partial}{\partial\sigma}\left(K_V\frac{\partial HS}{\partial\sigma}\right)-RS$$

Here, T and S stand for temperature and salinity, respectively. Heat balance and moisture balance at surface are considered as surface boundary condition for temperature and salinity, respectively. C_P and Q stands for specific heat coefficient and net surface heat flux at the surface, respectively. R stands for river discharge. k_H and k_V are the horizontal and vertical eddy diffusion coefficients, respectively.

3.1.2.3 Turbulence Closure

The vertical mixing coefficients, V_A and V_K are obtained by appealing to the second order turbulent closure scheme of Mellor and Yamada (1982), which characterizes turbulence by equations for the kinetic energy of turbulence, q^2, and turbulence macro scale l, according to:

$$\frac{Dq^2}{Dt}=K_q\frac{\partial q^2}{\partial z}+aA_v\left[\left(\frac{\partial u}{\partial z}\right)^2+\left(\frac{\partial v}{\partial z}\right)^2\right] \qquad (3.11)$$

$$+\frac{2g}{\rho_0}K_V\frac{\partial\rho}{\partial z}-\frac{2q^3}{B_1 l}+A_H\frac{\partial q^2}{\partial^2 x}+A_H\frac{\partial q^2}{\partial^2 y}$$

$$\frac{Dq^2}{Dt} = K_q \frac{\partial q^2}{\partial z} + aA_v \left[\left(\frac{\partial u}{\partial z} \right)^2 + \left(\frac{\partial v}{\partial z} \right)^2 \right] \tag{3.12}$$

$$+ \frac{2g}{\rho_0} K_V \frac{\partial \rho}{\partial z} - \frac{2q^3}{B_1 l} + A_H \frac{\partial q^2}{\partial^2 x} + A_H \frac{\partial q^2}{\partial^2 y}$$

Wall proximity functions $\overline{W}$ is defined as follows:

$$\overline{W} = 1 + E_2 \left(\frac{1}{kL} \right)^2 \tag{3.13}$$

where

$$\frac{1}{L} = \frac{1}{\eta - z} + \frac{1}{H + z} \tag{3.14}$$

Mixing coefficients are given as:

$$A_V = S_M l q \tag{3.15}$$

$$K_V = S_H l q \tag{3.16}$$

$$K_q = S_q l q \tag{3.17}$$

The stability functions S_M, S_H, and S_q are analytically derived from algebraic relations. They are functionally dependent on $\frac{\partial u}{\partial z}, \frac{\partial v}{\partial z}, g\rho_0 \frac{\partial \rho}{\partial z}$, q and l. These relations are derived from closure hypotheses described by Mellor (1973) and later summarized by Mellor and Yamada (1982).

3.1.2.4 Numerical Scheme

A semi-implicit finite difference scheme has been adopted where equations are discretized explicitly in the horizontal direction and implicitly in the vertical direction. An Arakawa C staggered grid has been used with the first order upwind scheme. The tri-diagonal formation of the momentum equation is utilized, and in combination with the mass conservation equation, an algebraic equation is obtained where the only unknown variable is the surface elevation η in implicit form. This algebraic equation is solved through the successive over relaxation (SOR) method.

3.1.3 Ecosystem Modeling of Coastal Regions

The phenomena in urban coastal zone are not only physical but also biological or chemical, each of which relates to the other. The ecosystems and water quality of urban coastal zones are highly complicated. To deal with these complex systems, a water quality model is composed of a three-dimensional physical circulation model and an ecosystem model that describes pelagic and benthic aspect of nutrients cycling. The pelagic and benthic systems also have interactions with each other.

In ecosystem models, each water quality variable is often called a compartment. Various kinds of models are also proposed for ecosystem models (Kremer and Nixon 1978; Fasham et al. 1990; Chai et al. 2002; Kishi et al. 2007). Some models, such as CE-QUAL-ICM (Cerco and cole 1995), Delft3D-WAQ (delft Hydraulics 2003), MIKE3_WQ [Danish Hydraulic Institute (DHI) 2005], RCA&ECCOM (Hydroqual 2004), etc., have been widely used to simulate water quality in estuaries and in the ocean.

Each model is developed with basic aquatic compartments such as phytoplankton, zooplankton, and nutrients. Some differences in the modeling of sediment, detritus, and the detailed modeling of phytoplankton exist, depending on the target ecosystems and the objectives of the study. As a result, no fully adaptive model applicable for all water bodies exists. If we were to make a model that could be adapted for all areas, its results would be too complex to discuss. It would not be so different from observing the real world. For example, ocean ecosystem models tend to focus only on pelagic systems. They tend to ignore benthic modeling, since the open ocean is deep enough to prevent the return of detritus to the seabed. Meanwhile, ocean ecosystem models generally deal with some metals in order to represent the limiting factor of phytoplankton. These metals are fully abundant in urban coastal zones. However, they are often depleted during phytoplankton growth in the open ocean. On the other hand, the concentration of phytoplankton in coastal areas is highly variable both spatially and temporally as compared to the open sea. Subsequent sedimentation of this bloom also constitutes a major input to benthic ecology (Waite et al. 1992; Matsukawa 1990; Yamaguchi et al. 1991). To represent these phenomena, ecosystem models of coastal zones usually cover benthic systems.

Ecosystem models solve conservation equations for relevant components with appropriate source and sink terms. This is the same as temperature and salinity modeling in physical models, as explained in Sect. 3.1.4.

For the sigma coordinate system, a mathematical formulation of the conservation of mass is written by:

$$\frac{\partial(HC)}{\partial t}+\frac{\partial(uHC)}{\partial x}+\frac{\partial(vHC)}{\partial y}+\frac{\partial(\dot{\sigma}HC)}{\partial\sigma}=$$

$$HK_H\left(\frac{\partial^2 C}{\partial x^2}+\frac{\partial^2 C}{\partial y^2}\right)+\frac{1}{H^2}\frac{\partial}{\partial\sigma}\left(K_V\frac{\partial HC}{\partial\sigma}\right)\pm S(x,y,\sigma,t)+W(x,y,\sigma,t)$$

where C denotes concentration of the water quality variable and t is time. Fluxes into and out of the target control volume are calculated by using physical model results. $S(x,y,\sigma,t)$ represents sources or sinks of the water quality variable due to internal production and the removal of the biogeochemical effect. It also represents the kinetic interactions of each compartment. $W(x,y,\sigma,t)$ represents the external inputs of the variable c.

For example, phytoplankton constitutes the first level in the food chain of the pelagic ecosystem of bays. Phytoplankton photosynthesizes by using sunlight and increases. At this time, the source term S of phytoplankton is increased depending on the amount of photosynthesis that takes place. Phytoplankton is then decreased by the grazing of zooplankton. The source term S of zooplankton is increased along with this grazing, and the source term of phytoplankton is decreased. Ecosystem models basically express the relationship of each compartment through mathematical expressions. These models provide a quantitative description of the influences of physical circulation on the biological and chemical processes of urban coastal zones.

3.1.3.1 Outline of an Ecosystem Model Description

The ecosystem model introduced here was developed to simulate the nutrient budget of an urban coastal zone. It includes the temporal and spatial variations of phytoplankton, nutrients, detritus, and dissolved oxygen (DO). In urban coastal zones, nutrients emitted from urban areas are not a limiting factor for phytoplankton growth. However, quantifying the nutrient budget is essential for analyzing and restoring the ecosystems of urban coastal zones. Fig. 3.4 shows schematic interactions of a lower trophic ecosystem model which is used for Tokyo Bay (Koibuchi et al. 2001) This model has 18 state variables: phytoplankton (Phy), zooplankton (Zoo), nutrients (NH4, NO3, PO4 and Si), labile detritus (LDON, LDOP, LDOSi) and refractory detritus (RDON, RDOP, RDOSi) for each nutrient, labile detritus carbon (LDOC), refractory detritus carbon (RDOC), dissolved organic carbon (DOC), and dissolved oxygen (DO), as well as sedimentation processed from particulate organic material.

Since the basic structure of the model follows the widely applied CE-QUAL-ICM (Cerco and Cole 1993, 1995), this section mainly focuses on our modifications of the CE-QUAL-ICM model in the following description.

3.1.3.2 Phytoplankton and Zooplankton Modeling

This model deals with four phytoplankton groups. Phyd1 is based on *Skeletonema costatum,* which is a dominant phytoplankton species in Tokyo Bay. Phyd2 represent a winter diatom group (such as *Eucampia*).

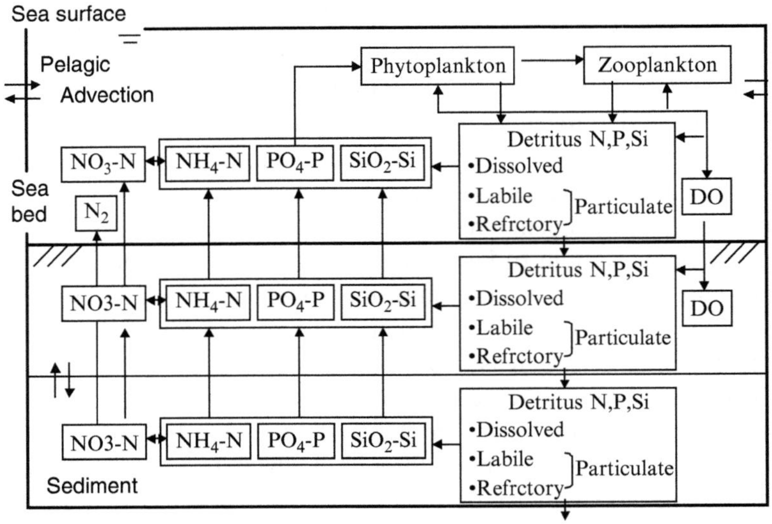

Fig.3-4. Idealized nutrient cycling in Tokyo Bay's ecosystem according to the model of Koibuchi et al. (2001). Cycling between the 18 state variables: phytoplankton, zooplankton, nutrients (nitrogen, phosphorus, and silicate), labile detritus and refractory detritus for each nutrient, and dissolved oxygen, as well as sedimentation processed from particulate organic materials

Phyr is a mixed summer assemblage consisting primarily of *Heterosigma akashiwo* and *Thalassosira*. Phyz denotes the dinoflagellates. These four phytoplankton assemblages have different optimal levels of light for photosynthesis, maximum growth rates, optimal temperatures for growth, and half saturation constants for nutrient uptake. Diatoms only use silica during growth.

The time rate of change of phytoplankton due to biological activity and sink is given by:

$$\frac{\partial Phy_x}{\partial t} = \mu_x Phy_x - gZoo - m_p Phy_x - W_{px}\frac{\partial Phy_x}{\partial z} \tag{3.18}$$

where $x = d1, d2, r, z$, denote each phytoplankton assemblage. The phytoplankton growth rate μ depends on temperature T, on photosynthetically available radiation I, and on the nutrient concentration of nitrogen, phosphorus, and silica :

$$\mu_x = \mu_{\max}(T) \cdot L(I) \cdot Min\left(L_N, L_P, L_{Si}\right) \tag{3.19}$$

$$\mu_{\max}(T) = \mu_{\max} \exp\left[-\beta_1 \left(T_{opt} - T\right)^2\right] \quad T \le T_{opt}, \tag{3.20}$$

or

$$\mu_{\max}(T) = \mu_{\max} \exp\left[-\beta_2 \left(T_{opt} - T\right)^2\right] T > T_{opt},$$

$$L(N) = (L_{NO3} + L_{NH4}) \tag{3.21}$$

$$L_{NO3} = \frac{NO3}{k_{NO3} + NO3} \cdot \frac{1}{1 + NH4 / k_{NH4}},$$

$$L_{NH4} = \frac{NH4}{k_{NH4} + NH4},$$

$$L(P) = \frac{PO4}{k_{PO4} + PO4} \tag{3.22}$$

$$L(Si) = \frac{Si}{k_{Si} + Si} \tag{3.23}$$

where $\mu_{\max}$ (T) is the growth rate at ambient temperature, which relates $\mu_{\max}$, the maximum growth rate $\mu_{\max} = \mu_0 \cdot 1.066^T$ (Eppey 1972), T_{opt}, the optimal temperature of each plankton assemblage, and β_1 and β_2 are shaping coefficients, k_{NO3}, k_{NH4}, k_{PO4} and k_{Si}, which is the Michaelis of the half-saturation constant for each nutrient. I is exponentially decreasing with water depth z according to:

$$I = I(z) = \tag{3.24}$$

$$I_0 \cdot par \cdot \exp\left\{-z\left(K_w + K_{\mathrm{chl}} \int_z^0 Ch(z)dz + K_{\mathrm{Sal}} \frac{1}{H+\eta} \int_z^0 Sal(z)dz\right)\right\}$$

where I_0 is shortwave radiation, and *par* is the fraction of light that is available for photosynthesis. K_w, K_{Chl}, and K_{Sal} are the light attenuation coefficients for water, chlorophyll, and depth average salinity, respectively.

Suspended sediment reduces underwater light intensity and affects the growth of phytoplankton. The effect of suspended sediment concentration on light intensity should be simulated as its own compartment. However, the re-suspension rate of mixed mud and the available data on the concentration of sediment suspended in river water and on the seabed are very limited. Therefore, we used observation results of salinity based on field observation data from 1999 and 2000, as shown in Fig. 3-5.

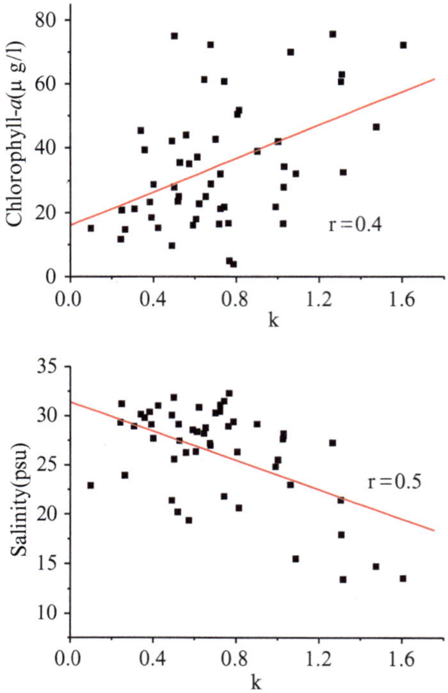

Fig. 3-5. Correlation between light attenuation constant k and Chlorophyll-*a* concentration (*top*), light attenuation constant k and salinity (*bottom*)

The function $L(I)$ represents the photosynthesis–light relationship (Evans and Parslow 1985),

$$L(I) = \frac{\alpha I}{\sqrt{\mu_{\max}^2 + \alpha^2 I^2}} \tag{3.25}$$

The rate of phytoplankton grazing, g, which is a function of an ambient temperature:

$$g = k_{grz} \theta_{grz}^{(T-20)} \tag{3.26}$$

where k_{grz} is the predation rate at 20°C. Other phytoplankton loss terms are mortality, represented by the linear rate m_p, where W_{px} is the constant vertical sinking velocity for each phytoplankton.

The growth rates of zooplankton are expressed as follows:

$$\frac{\partial Zoo}{\partial t} = g\beta Zoo - l_{BM} Zoo - l_E g\beta Zoo - m_Z Zoo^2 \tag{3.27}$$

Here β is the assimilation efficiency of phytoplankton by zooplankton, and l_{BM} and l_E denote excretion due to basal metabolism and ingestion, while the remaining fraction is transferred to the detritus. m_Z is the loss coefficient of zooplankton mortality.

3.1.3.3 Nutrients and Detritus Modeling

The nutrient compartments have four principal forms for each nutrient (nitrogen, phosphorus, and silica): dissolved organic nutrients, labile and refractory particulate organic nutrients (LPON and RPON, respectively), and dissolved inorganic nutrients. Only the dissolved inorganic nutrients are utilized by phytoplankton for growth. Nutrients are changed to these various organic and inorganic forms via respiration and predation. Fig. 3.6 shows an example of nutrient cycles using phosphorus. DOP, LPOP, and RPOP work as a pool of phosphorus.

For example, certain labile compounds that are rapidly degraded, such as the sugars and amino acids in the particulate organic matter deposited on the sediment surface, decompose readily; others, such as cellulose, are more refractory, or resistant to decomposition.

Table 3.1 shows the distributions of each detritus form by each event, based on Pett (1989).

Ammonia and nitrate are utilized by phytoplankton for growth. Ammonia is the preferred form of inorganic nitrogen for algal growth, but phytoplankton utilize nitrate when ammonia concentrations become depleted. Nitrogen is returned from algal biomass to the various dissolved and particulate organic nitrogen pools through respiration and predatory grazing. The time rates for variations due to the biological processes of nitrate and ammonium are as follows. Denitrification does not occur in the pelagic water systems in this model, but rather occurs in the anoxic sediment layer. As a result, if denitrification occurs in the sediment, nitrate is transferred by diffusion effect into the sediment.

$$\frac{\partial NO3}{\partial t} = -\mu_{\max\_X} L(I) L_{NO3} Phy_x + nNH4 \tag{3.28}$$

$$\frac{\partial NH4}{\partial t} = -\mu_{\max\_X} L(I) L_{NH4} Phy_x - nNH4 + l_{BM} Zoo + l_E g \beta Zoo + r_{DON} DON \tag{3.29}$$

Phosphorus kinetics is basically similar to nitrogen kinetics except for the denitrification and the alkaline phosphatase effects of the DOP degradation processes. Many phytoplankton can enhance alkaline phosphatase activity. This effect makes it possible for them to use phosphate from DOP pools

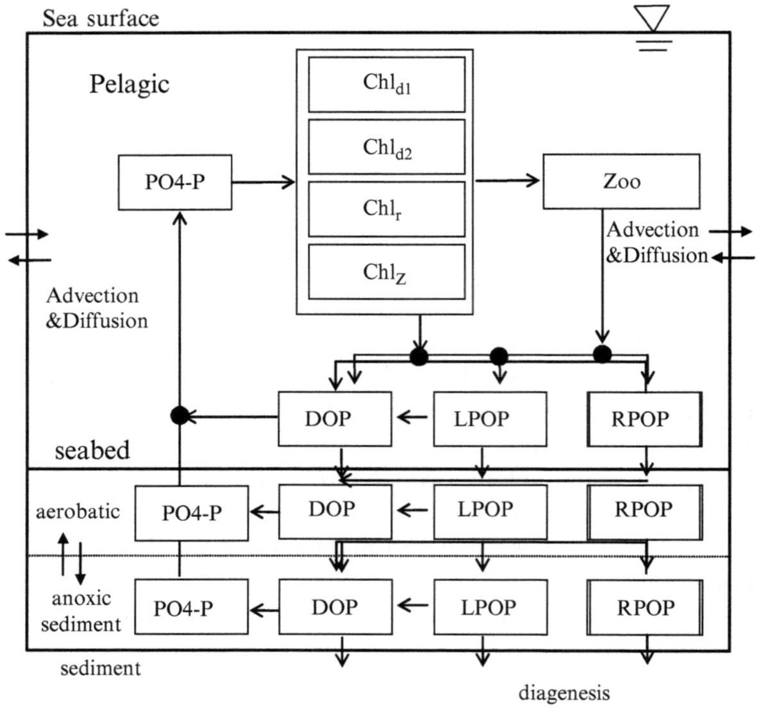

Fig. 3-6. Phosphorus cycles in the model

Table 3-1. Divide ratio of detritus

	Detritus type	Basal metabolism	Predation	Mortality of zooplankton
Carbon	*DOC* dissolved organic carbon	1	0.1	0.2
	LPOC labile particulate organic carbon	0		0.45
	RPOC refractory particulate organic carbon	0		0.35
Nitrogen	*DON* dissolved organic nitrogen	1	0.1	0.1
	LPON labile particulate organic nitrogen	0	0	0.45
	RPON refractory particulate organic nitrogen	0	0	0.35
	NH4 ammonia	0	0	0.1
Phosphorus	*DON* dissolved organic phosphorus	1	0.3	0.2
	LPON labile particulate organic phosphorus	0	0	0.35
	RPON refractory particulate organic phosphorus	0	0	0.15
	PO4 phosphate	0	0	0.3

(Fitzgerald and Nelson 1966). This effect is formulated in the following model:

$$\frac{\partial PO4}{\partial t} = -\mu_{\max\_X} L(I) L_{PO4} Phy_x + l_{BM} Zoo + l_E g \beta Zoo + r_{DOP} DOP \tag{3.30}$$

$$r_{DOP} = r_{DOP\_\min} + \frac{k_{PO4}}{k_{PO4} + PO4} r_{DOP\_di} \sum_{X=di1,di2} Chl_x \tag{3.31}$$

where, r_{DOP} is the decomposition rate for DOP, $r_{DOP\text{-}\min}$ is the minimum constant of DOP decomposition(day^{-1}), $k_{PO4}$is a half saturation constant of phosphate uptake, and $r_{DOP\text{-}di}$is the acceleration effect of DOP decomposition by diatoms.

The kinetics of the silica is fundamentally the same as the kinetics of the phosphorus. Only diatoms utilize silica during growth. Silica is returned to the unavailable silica pool during respiration and predation.

3.1.3.4 Sediment Processes

The sediment system is zoned in two layers (see Fig. 3.6), an aerobic and an anoxic layer. Organic carbon concentrations in the sediment are controlled by detritus burial velocity, the speed of labile and refrigerate organic carbon decomposition, and the rate constant for the diagenesis of particulate organic carbon. The thickness of the aerobic layer is calculated by oxygen diffusion when the amount of oxygen at the bottom layer of the pelagic system isn't zero. The nutrient model, which is a simplified version of the model, treats the nutrients ammonium, nitrate, phosphate, and silica and their exchanges with the pelagic system. Silicate-dependent diatoms and non-silicate-dependent algae are distinguished.

3.1.3.5 Dissolved Oxygen

Dissolved oxygen is an essential index for the water quality of an urban coastal zone. Sources of DO included in the model are reaeration at the sea surface, photosynthesis of phytoplankton, and DO in inflows. The sink of DO includes respiration of phytoplankton and zooplankton, oxidation of detritrial carbon (LDOC and RDOC), nitrification, and sediment oxygen demand. The time variation of DO is formulated as follows:

$$\begin{aligned}\frac{\partial DO}{\partial t} &= \mu_{\max\_X} Phy_x \cdot K_{OC} - m_p Phy_x \cdot K_{OC} - l_{BM} Zoo \cdot K_{OC} \\ &+ k_a \theta_a^{T-20} (DO_{sat} - DO) - k_{Nh4} \theta_{nh4}^{T-20} NH4 \frac{DO}{K_{nit} + DO} \cdot K_{ON} \\ &- K_{OC} \left(k_{RDOC} \theta_{rdoc}^{T-20} RDOC + k_{LDOC} \theta_{rdoc}^{T-20} LDOC \frac{LDOC}{K_{mLDOC} + LDOC} \right.\end{aligned} \tag{3.32}$$

where, K_{OC} is the oxygen to carbon ratio. K_{ON} is the oxygen to nitrogen ratio. k_a and θ_a^{T-20} denote the reaeration rate at 20°C and the temperature coefficient for reaeration at the sea surface, respectively. k_{Nh4} and θ_{nh4}^{T-20} are the ammonia oxidation rate at 20°C and the temperature coefficient. K_{nit} is the half saturation constant of ammonia oxidation. k_{RDOC} and θ_{rdoc}^{T-20} are the RDOC mineralization rate at 20°C and the temperature coefficient for RDOC mineralization. k_{LDOC} and θ_{Ldoc}^{T-20} mark the LDOC mineralization rate at 20°C and the temperature coefficient for LDOC mineralization. K_{mLDOC} is the half saturation constant for LDOC mineralization. The concentration of DO saturation is proportional to temperature and salinity. Oxygen saturation value is calculated by using the following equation:

$$DO_{Sat} = 14.6244 - 0.36713T + 0.0044972T^2 \\ -0.0966S + 0.00205ST + 0.0002739S^2 \tag{3.33}$$

where T is temperature and S is salinity.

3.1.4 Applications of Models to Nutrients Budget Quantifications

Tokyo Bay is located at the central part of the main island of Japan. The inner bay, the north of which is 50 km in length in its narrowest channel (Fig. 3.7) along the main axis of the bay, connects to the Pacific Ocean. Its average depth and width are 18 m and 25 km, respectively. Tokyo Bay is one of the most eutrophicated bays in Japan. Phytoplankton increase in the surface layer from late spring to early fall, and oxygen depletion and the formation of hydrogen sulfide occur on the sea bed. The sea-water color at the head of the bay sometimes becomes milky blue-green in late summer after a continuous north wind (Koibuchi et al. 2001). This phenomenon is called a blue tide.

In the last decade, a variety of water quality observation equipment has been developed. This has made it easier to measure water quality than in the past. However, even with advanced technology, measuring the flux of nutrients is not easy. To quantify the nutrients budget, we applied our numerical model to Tokyo Bay.

The computational domain was divided into 1km horizontal grids with 20 vertical layers. Computation was carried out from April 1, 1999 to March 31, 2000, with time increments of 300s provided by the Japan Meteorological Agency giving hourly meteorological data that included surface wind stress, precipitation, and solar radiation. At the open boundary, an observed tide level was obtained which can be downloaded from the Japan Oceanographic Data Center (JODC) of the Japan Coastal Guard. DON and DOP was obtained at 30% of TN, TP based on the observation results of Suzumura and Ogawa (2001) at the open boundary.

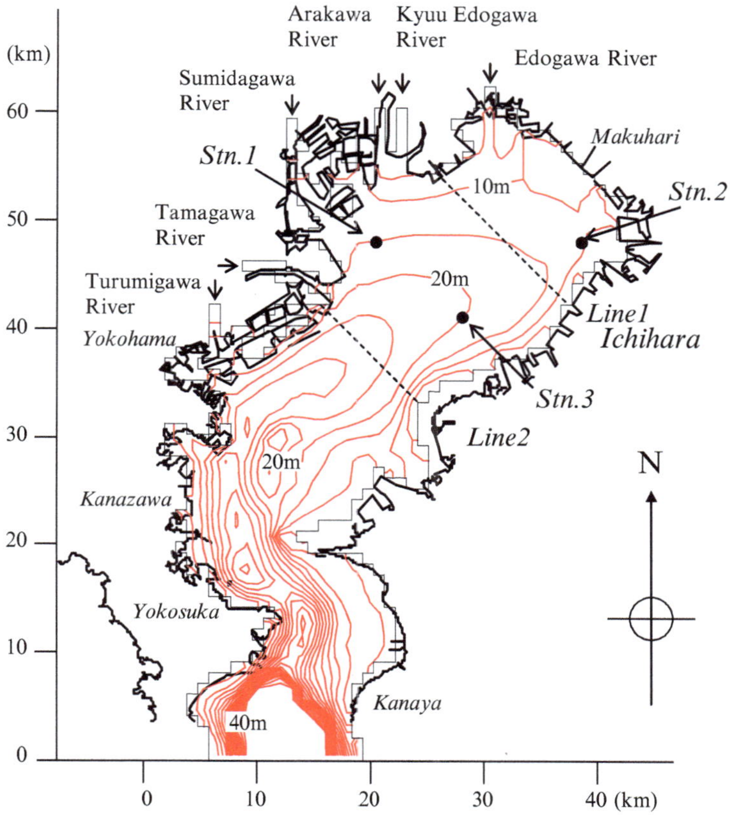

Fig. 3-7. The bathymetry of Tokyo Bay, Japan and its adjacent coastal area, showing the location of measurements in the bay

Fig. 3-8 shows a temporal variation of the computed density at S2. The simulation of water column density over the whole period (April 1999–October 1999) agreed well with measured density. Variations between simulated and observed values were generally less than 0.5 through the water column. Time variation of density effectively reproduced observed results, including short-term wind-induced variation, formation of stratification during summer, and mixing after continuous strong wind in the middle of October. Calculation results also reproduced an upwelling event during the summer season.

Total chlorophyll-a concentrations in the surface water were reproduced relatively well by model simulations. The model captured the temporal increase in chlorophyll-a that were seen in the observation results, as denoted by arrows in Fig. 3-9. During these periods, phytoplankton increased more than 50 μg/l. In Tokyo Bay, a red tide is defined as a chlorophyll *a* concentration of greater than 50μg/l. Four different types of plankton assemblages

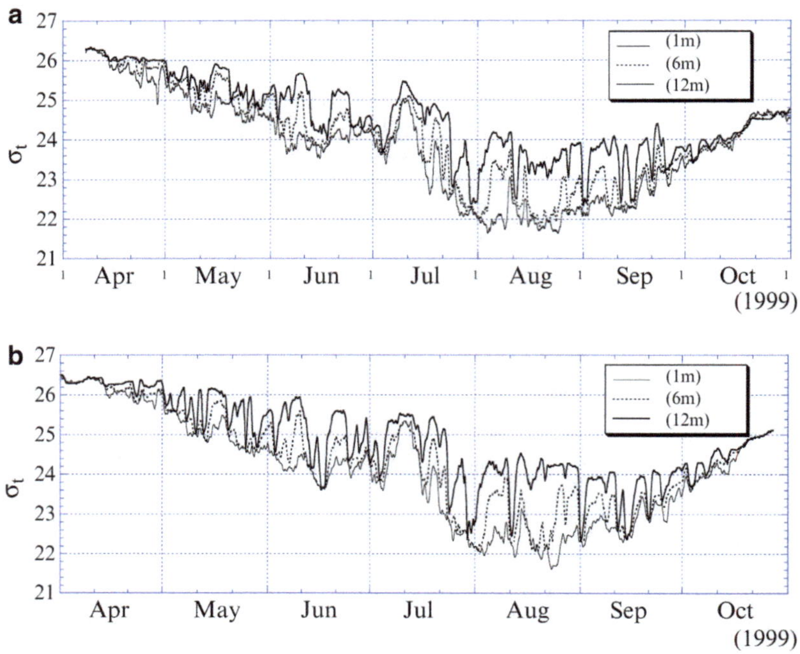

Fig. 3-8. Seasonal cycles of measured (top) and modeled (bottom) density

are defined in this model. This model underestimated the chlorophyll-a concentrations from late September through October.

DO showed high variability compared with field measurements at the bottom, and was relatively higher than the field data from late September through October. Oxygen-depleted water was made on the seabed, representing a basic trend for DO variations (Fig. 3-10).

The simulation of phosphate captured not only the observed increase in the surface layer during the summer season, but also inter-annual variability observed over the study period (Fig. 3-11). For example, phosphate concentrations increased from June to July due to phosphate release from sediment. Simulation results represented this kind of trend based on the oxygen-depleted water.

Fig. 3-12 shows nitrate concentration. Nitrate concentration in the surface layer fluctuated considerably during this period. Nitrate levels doubled or tripled occasionally at the surface. The timing of the high nitrate concentrations and the low density in the surface layer coincided with increases in the river discharge. Concentrations of nitrate were underestimated in bottom waters during summer. Further study is needed to simulate denitrification processes in the sediment layer.

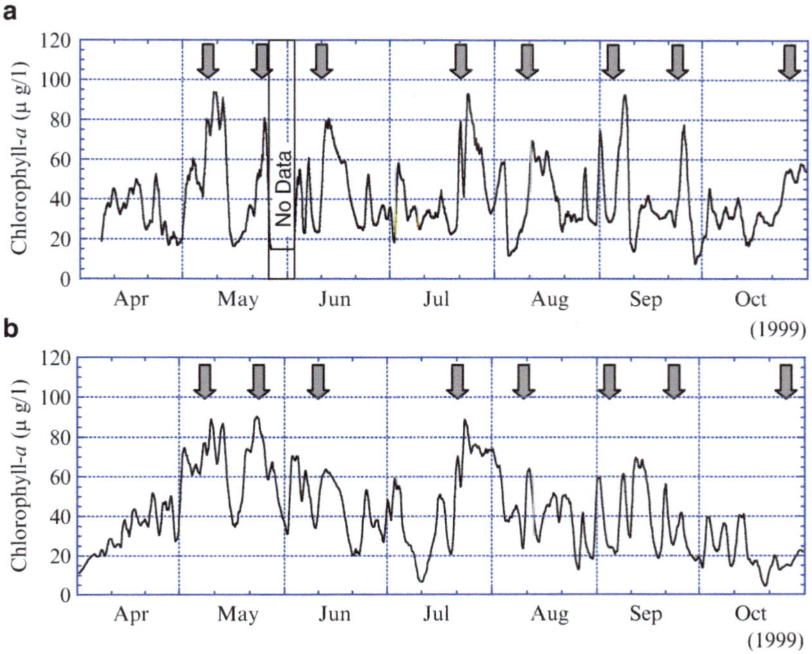

Fig. 3-9. Seasonal cycles of measured chlorophyll-*a* (top) and modeled chlorophyll-*a* (bottom). Arrows denote red tide events (over 50 μg/l)

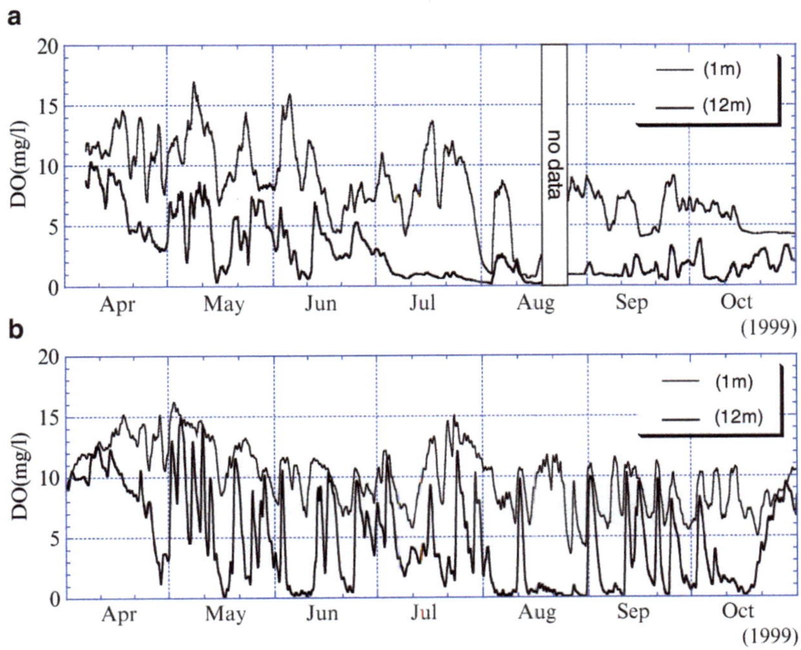

Fig. 3-10. Seasonal cycles of measured DO (top) and modeled DO (bottom) at 1 and 12 m from the sea surface

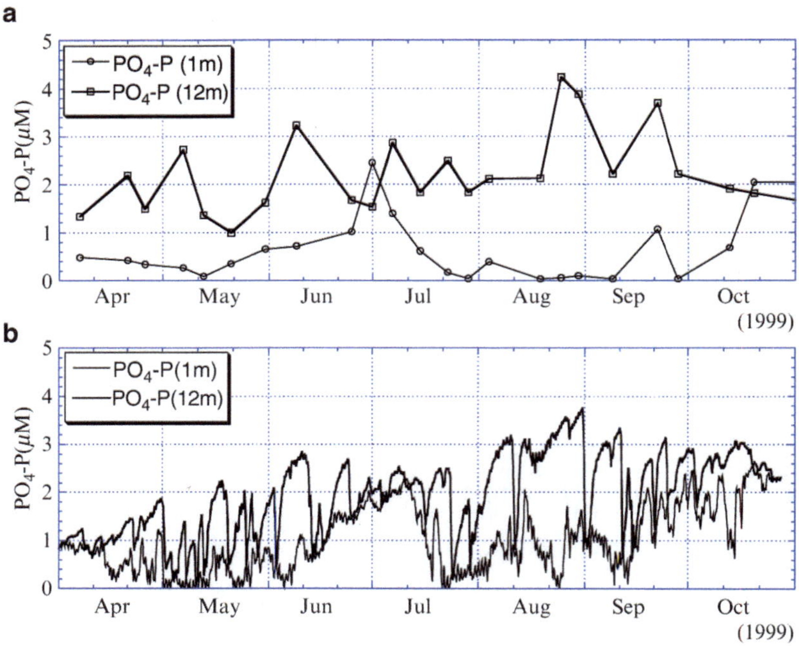

Fig. 3-11. Seasonal cycles of phosphorus measured (top) and modeled (bottom) at 1 and 12 m from the sea surface

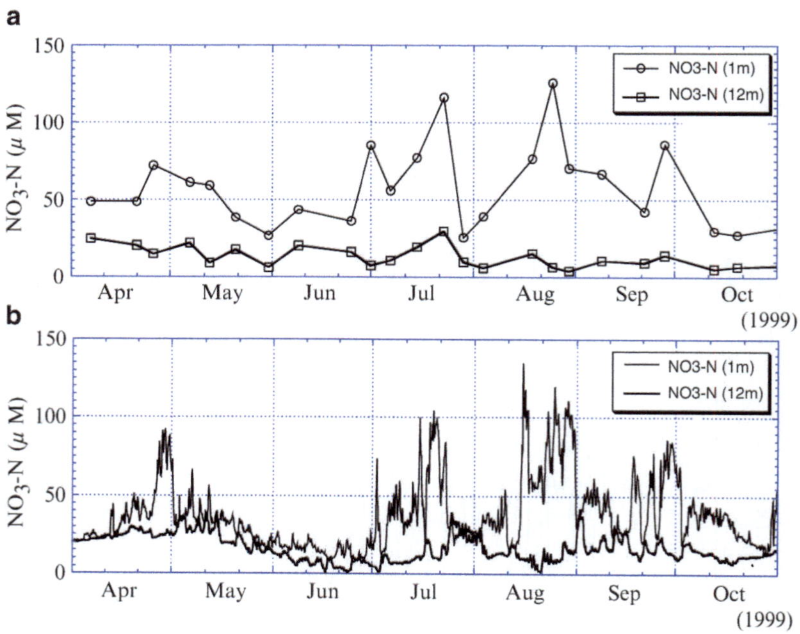

Fig. 3-12. Seasonal cycles of nitrate of measured (top) and modeled (bottom) at 1 and 12 m from the sea surface

Fig. 3-13 shows the calculation results of an annual budget of nitrogen and phosphorus in Tokyo Bay. The annual budget is useful in understanding nutrient cycles. Nitrogen is supplied to a considerable degree from rivers, since atmospheric nitrogen input is significant around urban areas. Phytoplankton uptake the nitrogen and sink to bottom waters, where they are decomposed by heterotrophic processes which consume oxygen. At the head of the bay (between the line 1 and line 2 in Fig. 3-7), about 40% of nitrogen is sunk as detritus and 20% of it is lost into the atmosphere by denitrification. Ammonia released from sediment reaches 20%. About 60% of the nitrogen load flows out from the bay. In contrast, atmospheric phosphorus input to the bay is negligible compared to the contribution of phosphorus from other sources.

Phosphate is released from sediment in the same amount as that discharged from the rivers, and it is transported to the head of the bay by estuarine circulation. As a result, the amount of phosphate in the inner bay remains high. In contrast, nitrogen is mainly supplied from the river mouth and transported quickly out of the bay. In conclusion, nitrogen and phosphorus showed important differences in the mechanisms by which they cycle in

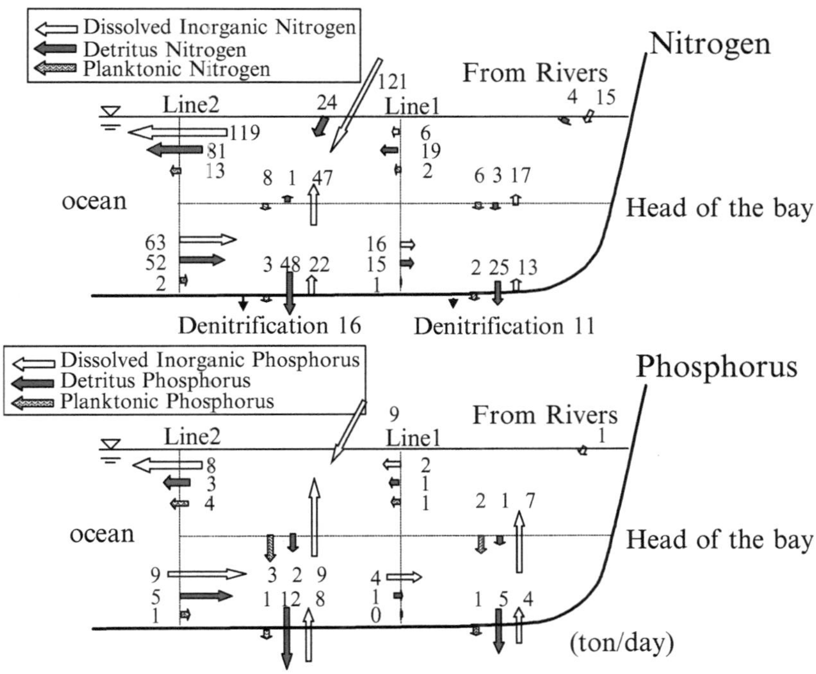

Fig. 3-13. Summary of fluxes and process rates calculated in Tokyo Bay from January 1999 to January 2000. Units are given in ton/year for each element

the bay. The regeneration of nutrients and their release from the sediment is an important source for phytoplankton growth and is equal to the contributions from the rivers. Phosphorus in particular is largely retained within the system through recycling between sediment and water. These results denote the difficulty of improving the eutrophication of bays through the construction of sewage treatment plants alone.

3.2 Application for CSO Modeling

3.2.1 Introduction

Big cities have long been developed near waterfronts. Even now, naval transport remains one of the most important transportation systems, especially for heavy industries and agriculture. Today, many of the world's largest cities are located on coastal zones, and therefore vast quantities of human waste are discharged into near-shore zones (Walker 1990). Fifty percent of the world's populations live within 100 km of the sea. Many people visit urban coastal zones for recreation and leisure, and we also consume seafood harvested from this area. As a result, effluents released into the water pose a risk of pathogen contamination and human disease. This risk is particularly heightened for waters that receive combined sewer overflows (CSOs) from urban cities where both sanitary and storm waters are conveyed in the same sewer system.

To decrease the risk from introduced pathogens, monitoring that is both well designed and routine is essential. However, even though a surprising number of pathogens have been reported in the sea, measuring these pathogens is difficult and time consuming – not least because such pathogens typically exist in a "viable but non-culturable" (VBNC) state.

In addition, the physical environments of urban coastal zones vary widely depending on time and location. Their complicated geographical features border both inland and outer oceans, and so both inland and outer oceans affect them. For example, tidal currents, which are a dominant phenomenon in this area, oscillate according to diurnal periods. Even if the emitted levels of pathogens were constant and we could monitor the levels of pathogen indicator organisms at the same place, they would fluctuate according to tidal periods. Density stratification also changes with the tides.

Consequently, the frequent measurement of pathogens is needed to discuss the risk pathogens pose in urban coastal zones. However, this kind of frequent monitoring appears to be impossible. To solve this conundrum and achieve an assessment of pathogen risk, we developed a set of numerical models that expand upon the models developed in Sect. 3.1 and that include a pathogens model coupled with a three-dimensional hydrodynamic model.

Section 3.2.2 deals with the distributions of pathogens in urban coastal zones. The pathogens model is explained in greater detail in Sect. 3.2.3. Section 3.2.4 deals with numerical experiments that help to understand the effects of appropriate countermeasures.

3.2.2 Distributions of Pathogens in Urban Coastal Zones

Figure 3-14 shows some typical density distributions patterns in urban coastal zones. The changing balance between tidal amplitudes and river discharge is responsible for the differences among these patterns. As tidal currents increase, the production of turbulent kinetic energy grows and can become the largest source of mixing in the shallow coastal waters. On the other hand, river-discharged water has a low density, creating a density difference between sea water and land-input water.

In salt-wedge estuaries (Fig. 3-14, top), river water is discharged into a small tidal-range sea. The strength of the tidal currents decreases relative to the river flow. This creates a vertical stratification of density. As a result, river water distributes like a veil over the sea's surface and moves seaward.

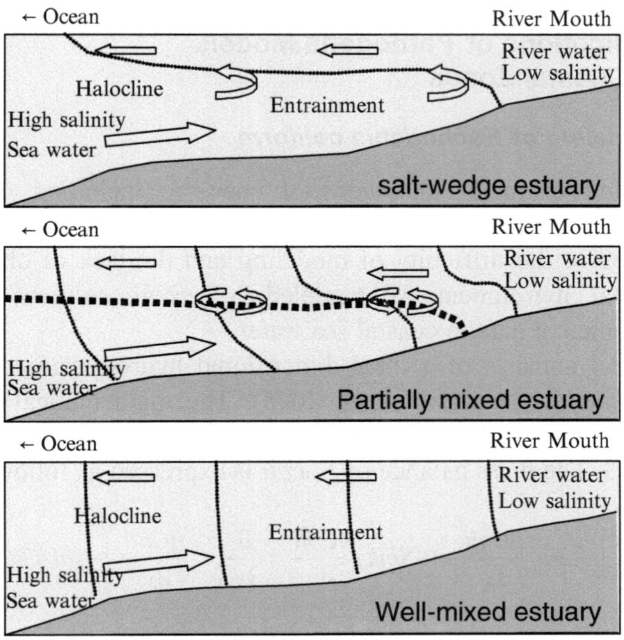

Fig. 3-14. Cross-sectional view of the mixing patterns in urban coastal zones

In contrast, bottom water moves to the river mouth and mixes with the river water. Under such conditions, pathogens move on the surface of the sea, and further mixing with low-density fresh water is restricted by stratification.

In partially mixed estuaries (Fig. 3-14, middle), the tidal force becomes a more effective mixing mechanism. Fresh water and sea water are mixed by turbulent energy. As a result, pathogens that are emitted from sewer treatment plants are more mixed than those in the static salt-wedge estuaries.

In well-mixed estuaries (Fig. 3-14, bottom), the mixing of salt and river waters becomes more complete due to the increased strength of tidal currents relative to river flow. Here, the density difference is developed in a horizontal direction. As a result, pathogens are mixed in the water column and settle down on the sea bed, in turn contaminating estuarine waters during the spring tide or contaminating rainfall through re-suspension (Pommepuy et al. 1992).

Figure 3-15 shows distributions of pathogens under coastal environments. These pathogens encounter a wide range of stresses including UV rays (Sinton et al. 2002), temperature differences (Matsumoto and Omura 1980), pH (Solić and Krstulović 1992), salinity (Omura et al. 1982), and lack of nutrients. The pathogens are transported by currents and continue to become part of sedimentation and to be re-suspended in urban coastal zones (Pommepuy et al. 1992).

3.2.3 Applications of Pathogens Models in Urban Coastal Zones

3.2.3.1 Modeling of Escherichia coliform

The modeling of major pathogens of concern (including Adenovirus, Enterovirus, Rotavirus, Norovirus, and Coronavirus) is not usually conducted owing to the difficulty of modeling and the lack of observational data in coastal environments. We modeled *Escherichia coliform (E. coli)* by using experimental data in coastal sea water.

This model consists of a three-dimensional hydrodynamic model and an *E. coli* model (Onozawa et al. 2005). The mathematical framework employed in the *E. coli* model takes the same approach that was explained in Sect. 3.1.3. The mass balance of *E. coli* is expressed as follows:

$$\frac{\partial Coli}{\partial t} + u_i \frac{\partial Coli}{\partial x_i} + (-Sink)\frac{\partial Coli}{\partial z} = \frac{\partial}{\partial x_i}\left(\varepsilon_i \frac{\partial Coli}{\partial x_i}\right) - sal \cdot Coli \qquad (3.33)$$

where *Coli* denotes concentrations of *E. coli* (CFU/100 ml), and t is time. *Sink* represents the sinking speed of *E. coli*. u_i denotes flow speed for the calculation of the advection term. ε_i denotes the diffusion coefficients. *sal*

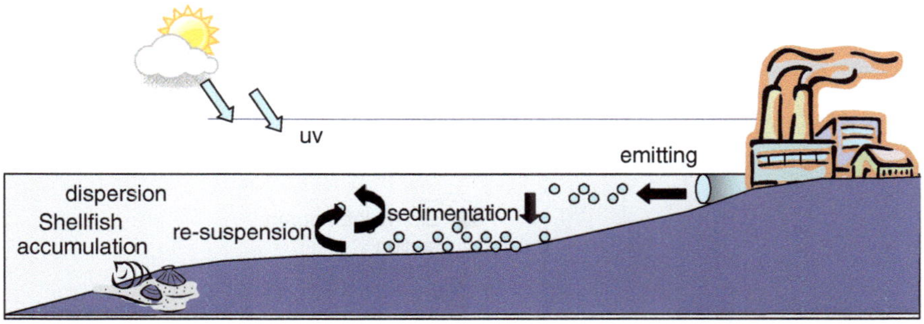

Fig. 3-15. Pathogens transportation in urban coastal zones

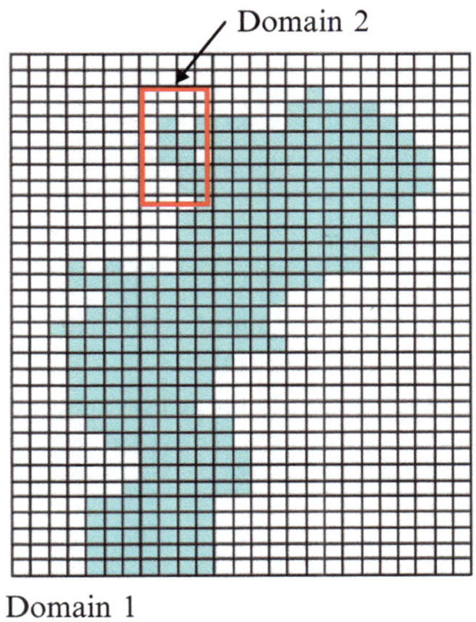

Fig. 3-16. Two nested computational domains

denotes the salinity-dependent die-off rate (ppt/day). Sunlight is generally recognized to be one source by which bacteria are inactivated, due to UV damage to the bacterial cell (Sinton et al. 2002). However, this particular target area has high turbidity that rapidly absorbs UV rays at the sea's surface. As a result, this process has been ignored in this model.

In this model, we can see the numerical simulations performed with two nested domains to fit the complex geography feature around the Odaiba area (Figs. 3-16 and 3-17). These nested grids make possible a representation of the stratification effect. A detailed configuration of the model is summarized in Table 3-2. The two computational domains cover the whole

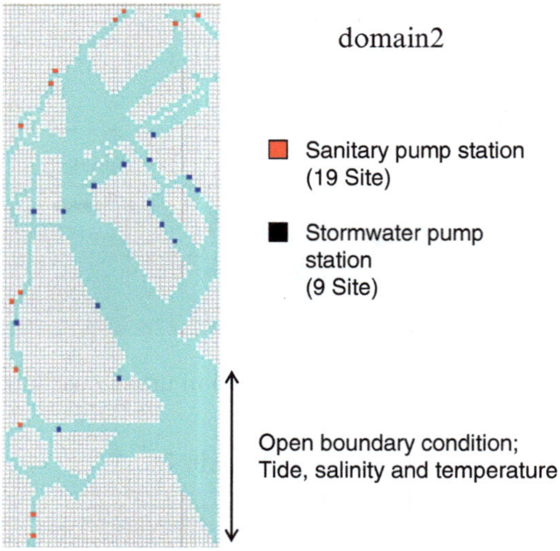

Fig. 3-17. Computational domain 2 and pumping station distributions around the Odaiba area

region of Tokyo Bay and the Odaiba area with grid resolutions of 2 km and 100 m, respectively. The first domain size is 25 × 33 grid points and the second domain has 50 × 127 grid points. All of the domains have 10 vertical sigma levels.

3.2.3.2 Model Validation

Figure 3-18 shows a Tokyo Bay scale grid (domain 1 in Fig. 3-16) that includes both the salinity simulation and observation results. The simulation results show stratification, mixing, and an upwelling phenomenon, and include levels and timing. Fig. 3-19 shows a comparison between observation results and a calculation for temperature and salinity in a fine grid scale (domain 2 in Fig. 3-16). Variations between the simulated and observed values were generally less than 2.5°C and 2 psu through the water column. The timing and periods of upwelling events were captured accurately. After rain fall, river discharge was increased remarkably. Model results adequately represent precipitation variation events and their effects.

Figure 3-20 shows a comparison between modeled and measured *E. coli* at Stn.1. The current standard for acceptably safe beaches for swimming set by of the Ministry of the Environment of Japan is a fecal coliform rate of 1000 coliforms unit per 100 mL (CFU/100 mL). This index of fecal coliform includes not only *E. coli* but also others. However, it is well known that

Table 3-2. Boundary conditions and grid resolution

	Domain 1	Domain 2
Computational area	50 km × 66km	5 km × 12.7 km
Grid size (m)	2,000	100
Number of grid	25 × 33 × 10hyer = 8,250	50 × 127 × 10hyer = 63,500
Total number of grid	2,920	22,740
Computational duration	2004/April/1 ~October/31	2004/August/1 ~October/15
Time step	10 min	30 s
Total time step	30,240 step	218,880 step

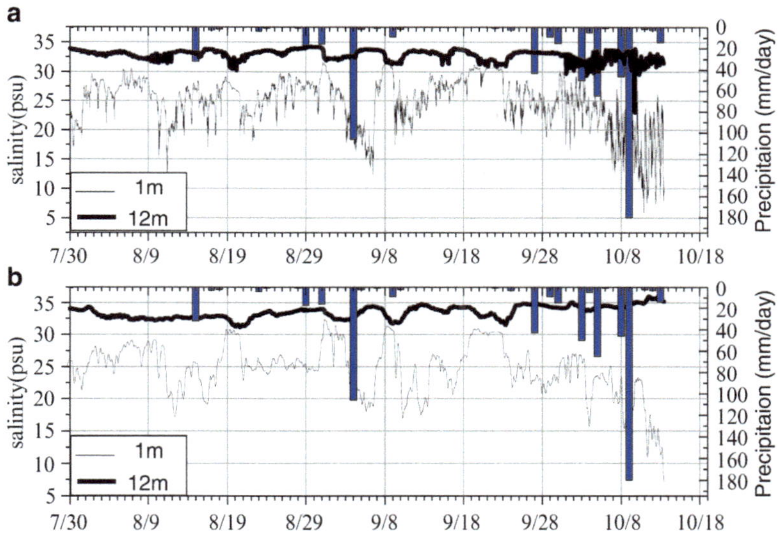

Fig. 3-18. Surface (depth 1 m) and bottom (depth 12 m) salinity derived from observation (**a**) and model simulations (**b**) for the period July 30–October 15

the majority of the fecal coliform in this area comes from *E. coli*. Therefore, we use a value of 1000 CFU/100 ml *E. coli* as the standard for the safety of swimming in the sea. From the calculation results, we can see that durations when the standards are exceeded are very limited, and that most of the summer period falls below the standard for swimming. This result also denotes that the increasing rates of *E. coli* do not agree with levels of precipitation. Even in small precipitations, *E. coli* significantly increased.

Understanding the effects of physical factors is important to understanding the fate and distributions of pathogens. Such an understanding is in turn

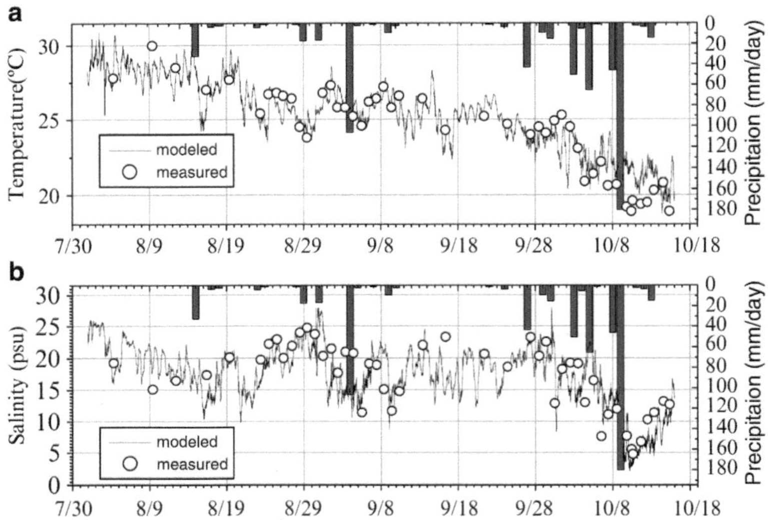

Fig. 3-19. Comparison between measured and modeled temperature (**a**) and salinity (**b**) at Stn.3 with precipitation

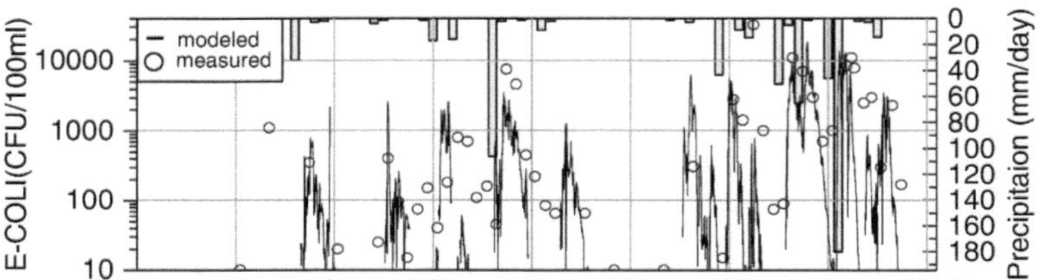

Fig. 3-20. Comparison between measured and modeled *E. coli* at Stn.1

highly related to the assessment of sanitary risks in urban coastal zones. Numerical experiments were thus conducted to examine the variations in rates of *E.coli* according to time and space.

3.2.4 Numerical Experiments with the Pathogens Model in Urban Coastal Zones

3.2.4.1 Numerical Experiments

Variations in levels of *E. coli* are directly correlated with the discharge from pumping stations, tidal currents, river discharges, and density distributions, as explained in Sect. 3.2.2. As a result, the distributions of CSO differ according to timing, even when the level of discharge is the same. We performed numerical experiments in order to evaluate the contributions of these different discharges and phenomena to CSO distributions. The first numerical experiment was a nowcast simulation that calculated *E. coli* distributions under realistic conditions. The second experiment was a numerical experiment to estimate the effects of a waste-reservoir that was being constructed near the Shibaura area. Numerical experiments were also applied Odaiba area, which is used as a bathing area.

3.2.4.2 Nowcast Simulations of E. coli Distributions

Figure 3-21 shows temporal variations of precipitation and river discharges (top), as well as tide levels and *E. coli* concentrations discharged from three different areas. Shibaura and Sunamachi area are located at the upper bay location from the Odaiba area. Morigasaki has the largest area, but is located in the lower bay location from the Odaiba area (see Fig. 3-17).

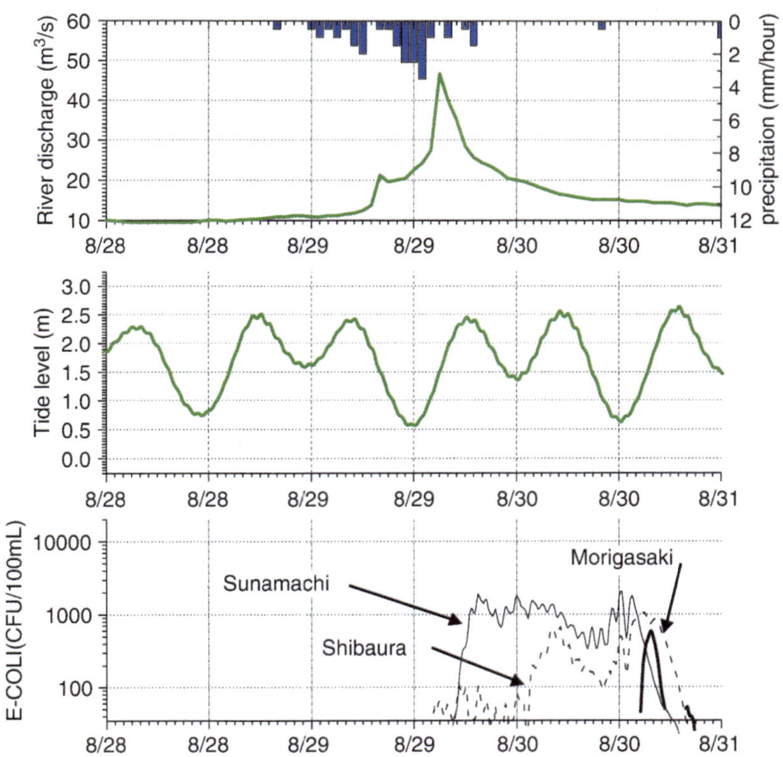

Fig. 3-21. Effect of rainfall and tides for *E. coli* variations under the small precipitation case

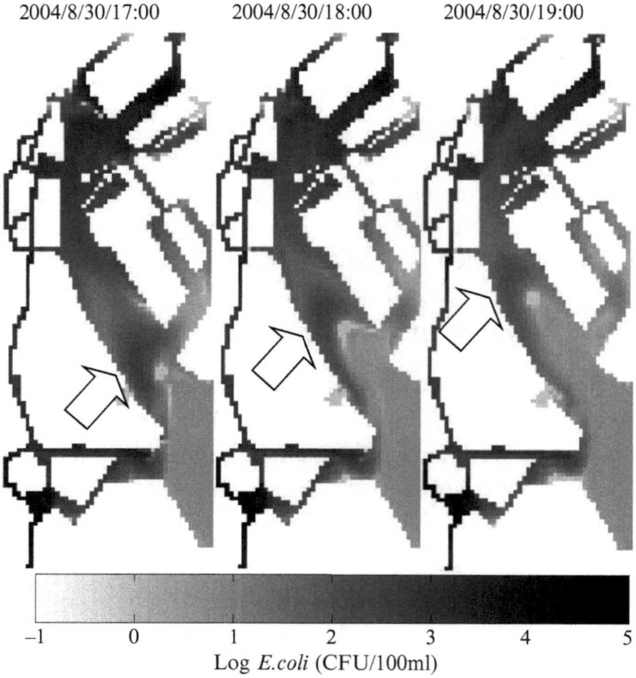

Fig. 3-22. Spatial distributions of *E. coli* at Odaiba area

In this spring tide period, tidal ranges can reach 2 m. Small precipitations were measured from August 29th to 30th. River discharge increased with precipitation, reaching 50 m^3/s. The levels of *E. coli* increased rapidly after the rainfall event, due mainly to discharges from the Sunamachi and Chibaura areas. Near the end of this period of increase, effluent from Morigasaki also reached the Odaiba area. Fig. 3-22 shows the spatial distributions of *E. coli* from three different times. From this Fig., the *E. coli* emitted from the Morigasaki area can be seen to have been transported from the lower region of the bay to the Odaiba area. This is because the small amount of river discharge resulted in a thin layer of low-density, highly concentrated *E. coli* on the surface of the sea, and tended to isolate the *E. coli* by preventing it from mixing with the water column.

In contrast, Fig. 3-23 shows a large precipitation case under the neap tide period. Large amounts of precipitation produced a large river discharge that reached 500 m^3/s. In this period, only the upper bay's CSOs arrived at the Odaiba area. No contributions from the Morigasaki area took place.

In conclusion, the concentrations of *E. coli* vary widely according to space and time. The density distributions produced by the balance of

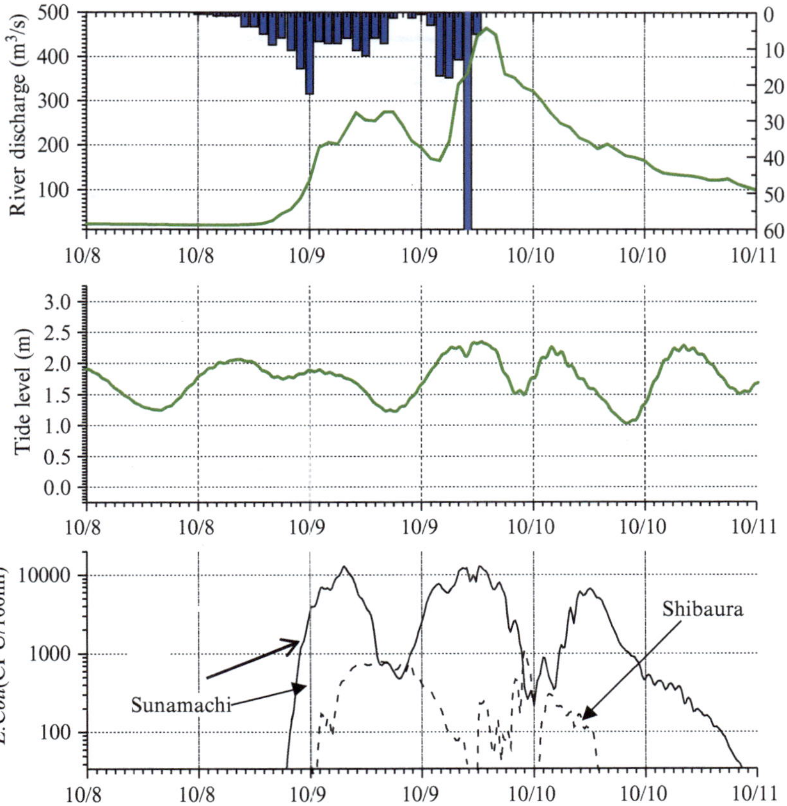

Fig. 3-23. Effect of rainfall and tides for *E. coli* variations under the large precipitation case

tides and river discharges have very complex effects. *E. coli* concentrations reached maximum levels after small precipitation events, but did not increase so much under large precipitation events due to mixing. These kinds of results would be impossible to understand only from observation. The model successfully captured complex distributions of *E. coli* and helped our understanding of pathogens contaminations.

3.2.4.3 Improvement of CSO Systems Through the Construction of Storage Tanks

To mitigate CSO pollutions, the construction of storage tanks at three sites in Tokyo has been planned by the Tokyo Metropolitan Government. Shibaura is the target area of this plan around the Odaiba area. Numerical

Table 3-3. Effects of storage tank

	Stn.l	Stn.2	Stn.3	Stn.4	Stn.5
Before	8.21	2.06	15.44	10.48	3.6
After	6.35	1.96	11.42	8.53	3.31
Improvement	1.86	0.1	4.02	1.95	0.29

Unit: day

experiments were conducted to evaluate the effects of storage tanks on CSOs. Calculation results were compared by the duration of periods when the amount of CSOs in the Odaiba area exceeded bathing standards.

Numerical simulation was performed with and without the proposed storage tank, which has a capacity of 30,000 m^3. This storage tank can store CSOs after rainfall. To include the effect of continuous rain, we assumed that the CSOs stored in the tank could be purified within 1 day.

Table 3.3 shows the calculation results for the mitigation effect of the storage tank. These numbers denote the dates when the standards for bathing in the sea (over 1000 CFU/100 ml) were exceeded. Fig. 3.17 shows the observation stations. From this table, Stn.3 shows the largest decrease in CSOs among these five stations. This is because Stn.3 is located closest to the pumping stations, and therefore would be most sensitive to the CSOs. Before the construction of the CSO storage tank, minimum bathing standards for CSO levels were exceeded on 15 days. After the construction of the storage tank, the duration of CSO levels that exceeded safety standards decreased to 11 days. There was an improvement of 4 days. On the other hand, other stations only 2 days, or in some cases, less than 1 day. Such differences could not be observed, especially in those stations that are located inside the Odaiba area due to the enclosed feature of bathymetry.

For example, over 10,000 storage tanks have been built in Germany alone, and another 10,000 were planned during the 1980s in Germany. Our plans for dealing with CSOs are not enough to mitigate the effects of CSOs completely. At the same time, these results show us the complexity of pathogens distributions and the importance of numerical modeling for this problem.

3.2.5 Summary

Numerical simulation is one of the most important tools for the management of water quality and ecosystems in urban coastal zones. We have developed a water quality model to simulate both nutrient cycles and pathogens distributions, and coupled it with a three-dimensional hydrodynamic model of urban coastal areas. To quantify the nutrients budget, a numerical model should include material cycles with phytoplankton, zooplankton, carbons, nutrients, and oxygen. We applied this model to the Tokyo Bay and simulated water column temperatures, salinity, and nutrient concentrations that were closely linked with field observations. This model successfully captured periods of timing, stratification events, and subsequent changes in bottom water oxygen and nutrients. Our model results also indicated that there were clear differences between the material cycles of nitrogen and phosphorus inside the bay. The regeneration of nutrients and its release from sediment was found to be a source of phytoplankton growth on the same order of importance as contributions from rivers. In particular, phosphorus was found to have been largely retained within the system through recycling between sediment and water.

We also developed a pathogen model that includes *E. coli* and is applied to the simulation of CSO influences in urban coastal zones. These results indicate that, because of stratification, concentrations of *E. coli* significantly increase after even small precipitation events. From this study, the balance between tidal mixing and river waters can be seen to be significant. However, these are only two case studies; it remains necessary to simulate the structure and characteristics of CSO distributions and their impact on urban coastal zone pollution. Such simulations remain as future works to be undertaken.

References

Ærtebjerg G, Andersen JH, Hansen OS (2003) Nutrients and Eutrophication in Danish Marine Waters, Vol. Ministry of the Environment, National Environmental Research Institute

Blumberg AF, Mellor GL (1987) A description of a three-dimensional coastal ocean circulation model, Vol. American Geophysical Union, Washington, DC.

Caperon J, Cattel SA, Krasnick G (1971) Phytoplankton kinetics in a subtrophical estuary: eutrophication. Limnol Oceanogr 16:599–607

Cerco CF, Cole T (1993) Three-dimensional eutrophication model of Chesapeake Bay. J Environ Eng 119:1006–1025

Cerco CF, Cole T (1995) User's guide to the CE-QUAL-ICM: three-dimensional eutrophication model, Vol. Vicksburg, MS

Chai F, Dugdale RC, Peng TH, Wilkerson FP, Barber RT (2002) One Dimensional Ecosystem Model of the Equatorial Pacific Upwelling System Part I: Model Development and Silicon and Nitrogen Cycle. Deep-Sea Res II 49:2713–2745

Eppley RW (1972) Temperature and phytoplankton growth in the sea. Fish Bull 70:1063–1085

Evans GT, Parslow JS (1985) A model of annual plankton cycles. Biol Oceanogr 3:327–347

Fasham MJR, Ducklow HW, Mckelvie SM (1990) A nitrogen-based model of phytoplankton dynamics in the oceanic mixed layer. J Mar Res 48:591–639
Fitzgerald GP, Nelson TC (1966) Extractive and enzymatic analysis for limiting or surplus phosphorus in algae. J Phycol 2:32–37
Freeman NG, Hale AM, Danard MB (1972) A modified sigma equations; approach to the numerical modeling of great lake hydrodynamics. J Geophys Res 77: 1050–1060
Haidvogel DB, Arango HG, Hedstrom K, Beckmann A, Rizzoli PM, Shchepetkin AF (2000) Model evaluation experiments in the North Atlantic Basin: simulations in nonlinear terrain-following coordinates. Dyn. Atmos. Oceans 32:239–281
Hydroqual (2004) User's guide for RCA, Release 3.0., Hydroqual, Inc., NJ
Jøgensen BB, Richardson K (1996) Eutrophication in Coastal Marine Ecosystem, vol 52. American Geophysical Union, Washinton DC
Johnson BH, Kim KW, Heath RE, Hseish NN, Butler HL (1993) Verification of a three-dimensional hydrodynamic model of Chesapeake Bay. J Hydraul Eng 119:2–20
Kantha LH, Clayson CA (2000) Small Scale Processes in Geophysical Fluid Flows, Academic Press. 888pp
Kishi MJ, Kashiwai M, Wared DM, Megreye BA, Eslingerf DL, Wernerg FE, Maki Noguchi-Aitab TA, Masahiko Fujii j w, Shinji Hashimotok, Daji Huangl, Hitoshi Iizumim, Yukimasa Ishidav, Sukyung Kango GAK, Hyun-cheol Kimo, Kosei Komatsun, Vadim V. Navrotskyq SLS, Kazuaki Tadokorob,x, Atsushi Tsudam,r, Orio Yamamuram YY, b, Katsumi Yokouchis, Naoki Yoshiei,v, Jing Zhangt YIZ, Vladimir I. Zvalinskyq (2007) NEMURO – a lower trophic level model for the North Pacific marine ecosystem. Ecol Modell 202:12–25
Koibuchi Y, Isobe M (2005) Blue Tide occurred in the west of Tokyo Bay in Summer of 2004. Proceeding 3rd international conference on Asia and Pacific Coast 1512-1521
Koibuchi Y, Sasaki J, Isobe M (2001) Study on Budget and Circulation of Nitrogen and Phosphorus in Tokyo Bay. Proc Coast Eng 48:1076–1080
Kremer JN, Nixon SW (1978) A coastal marine ecosystem: simulation and analysis, vol 24. Springer, Heidelberg
Li M, Zhong L, Boicourt WC (2005) Simulations of Chesapeake Bay estuary: sensitivity to turbulence mixing parameterizations and comparison with observations. J Geophys Res 110:C12004. doi: 10.1029/2004JC002585
Lucas LV, Koseff JR, Colern JE, Monismith SG, Thompson JK (1999) Processes governing phytoplankton blooms in estuaries. II. The role of transport in global dynamics. Mar Ecol Prog Ser 187:17–30
MacCready P, Hetland RD, Geyer WR (2002) Long-term isohaline salt balance in an estuary. Cont Shelf Res 22:1591–1601
Matsukawa YKS (1990) Nitrogen budget in Tokyo Bay with special reference to the low sedimentation to supply ratio. J Oceanogr 46:44–54
Matsumoto J, Omura T (1980) Some factors affecting the survival of fecal indicator bacteria in sea water. Technol Rep 45:169–185

Mellor GL (1973) Analytic prediction of the properties of stratified planetary surface layers. J Atmos Sci 30:1061–1069

Mellor GL, Yamada T (1982) Development of a turbulence closure model for geophysical fluid problems. Rev Geophys 20:851–875

Omura T, Onuma M, Hashimoto Y (1982) Viability and adaptability of E-COLI. and Enterococcus group to salt water with high concentration of sodium chloride. Wat Sci Tech 14:115–126

Onozawa K, Koibuchik Y, Furumai H, Katayama H, Isobe M (2005) Numerical calculation of combined sewer overflow(CSO) due to heavy rain around Daiba in the head of Tokyo Bay. Annu J Coast Eng 52:891–895

Pett R (1989) Kenetics of microbial mineralization of organic carbon from detrital *Skeletonema costatum* cells. Mar Ecol Prog Ser 52:123–128

Phillips NA (1957) A coordinate system having some special advantages for numerical forecasting. J Meteorol 14:184–185

Pommepuy M, Guillaud JF, Dupray E, Derrien A, Le Guyader F, Cormier M (1992) Enteric bacterial survival factors. Water Sci Technol 25:93–103

Prudman J (1953) Dynamical oceanography, Vol. Methuen & Co, London

Sinton LW, Hall CH, Lynch PA, Davies-Colley RJ (2002) Sunlight inactication of fecal indicator bacteria and bacteriophages from waste stabilization pond effluent in fresh and saline waters. Appl Environ Microbiol 68:1122–1131

Solić M, Krstulović N (1992) Separate and combined effects of solar radiation, temperature, salinity, and pH on the survival of feacal coliforms in seawater. Mar Pollut Bull 24:411–416

Suzumura M, Ishikawa K, Ogawa H (1998) Characterization of dissolved organic phosphrous in coastal seawater using ultrafiltration and phosphoydrolytic enzymes. Limnology and Oceanography 43:1553–1564

Unoki S (1998) Relation between the Transport of Gravitational Circulation and the River Discharge in Bays. J Oceanogr 7:283–292

Waite A, Bienfang PK, Harrison PJ (1992) Spring bloom sedimentation in a subarctic ecosystem. Mar Biol 114:131–138

Walker HJ (1990) The coastal zone. In the Earth as Transformed by Human Action. Local and Regional Changes in the Biosphere over the Past 300 Years, Vol. Cambridge University Press, Cambridge

Yamaguchi Y, Satoh H, Aruga Y (1991) Seasonal changes of organic carbon and nitrogen production by phytoplankton in the estuary of River Tamagawa. Mar Pollut Bull 23:723–725

4. Analysis of Natural Cross-Ventilation for Building Environmental Control

Motoyasu Kamata, Masashi Imano, Yoshihiko Akamine, Yunchan Zheng, Hideaki Hoshino, and Yu-Feng Tu

4.1 Utilization of Cross-Ventilation in the High-Density Urban Area

4.1.1 Necessity of the Research on Cross-Ventilation Utility and Its Current Situation

An ancient Japanese well-known essayist, named Kenko, a Buddhist priest, wrote the "Essays in Idleness," which is called "Tsurezure-gusa" in Japanese. Section 55 of these essays says, " A house should be built with the summer in mind. In winter it is possible to live anywhere, but a badly made house is unbearable when it gets hot." Even in modern days, this section is often referred to since it eloquently describes the unbearable hot and humid climate of Japanese summer as well as the Japanese housing conditions. In short, the Japanese traditional houses had such characteristics as having long eaves to prevent sunshine of summer from coming into rooms, or being equipped with large open passages with sliding-wooden-doors for cross-ventilation utilization.

However, there is less outdoor wind blowing into the building due to increased density of buildings in urban areas in Japan and Southeast Asia over the past several years. As a result, there is no sufficient amount of wind in the indoor. Additionally, due to concerns of security and privacy, houses with many or large size of openings are difficulty to be built. Despite the above mentioned factors that discourage utilization of cross-ventilation, residents still desire to have more natural airflow in the house based on the results from questionnaire survey (see Sects. 4.2 and 4.3 for details). Therefore, it is important to develop a design that improves amount of airflow. Shiraishi et al. (2002) demonstrated that the cross-ventilation flow

H. Furumai et al. (eds.), *Advanced Monitoring and Numerical Analysis of Coastal Water and Urban Air Environment*,

© 2010 to the complete printed work by Springer, except as noted. Individual authors or their assignees retain rights to their respective contributions; reproduced by permission.

rate increases and cooling load is significantly reduced in the Porous buildings. Narumi et al. (Narumi et al. 2007; Ikenoue et al. 2002; Habara et al. 2002) also showed the cross-ventilation effect of monitor roof by conducing real house experiments, wind tunnel experiment, and computational fluid dynamics.

4.1.2 Wind Pressure Characteristics of Detached House in the High-Density Urban Area

The current results were obtained from wind tunnel experiments conducted in the Wind Environment Simulator laboratory at the University of Tokyo. Fig. 4-1 shows the airflows of wind tunnel. Velocity at eaves height (59 mm) of target house model (the detached house) was set as 7 m/s; the exponent of the velocity profile was set as 0.2 under the assumption of being in residential area. Fig. 4-2 shows the target model used in the benchmark test of heat load simulation program, which was developed by the research committee on Thermal Environmental Engineering of Architectural Institute in Japan. Scale of the target model was set as 1/100 and the roof gradient was ½. There were 126 pressure measurement points on the model. The study examined the value differences measured from these pressure measurement points and static pressure of the Pitot tube, which was placed in the wind tunnel.

Wind pressure coefficients C_p were calculated via Eq. 4.1 presented below. P_d (standard dynamic pressure) indicates the dynamic pressure measured at the eaves height without the existence of any models. P_s indicates the static pressure measured from the Pitot tube; P_w indicates the wind pressure measured from the measurement points on the model.

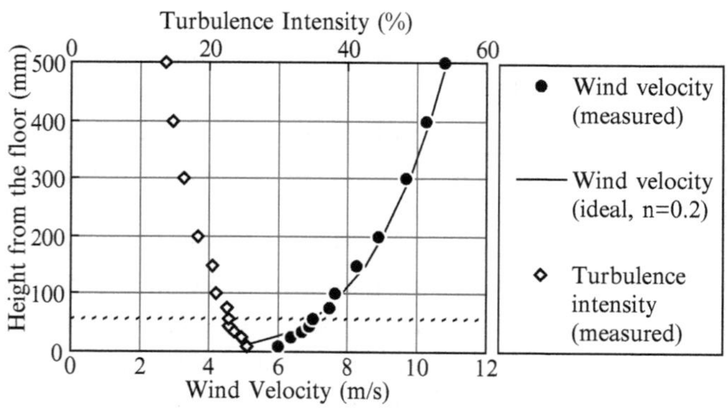

Fig. 4-1. The airflow-profile of wind tunnel

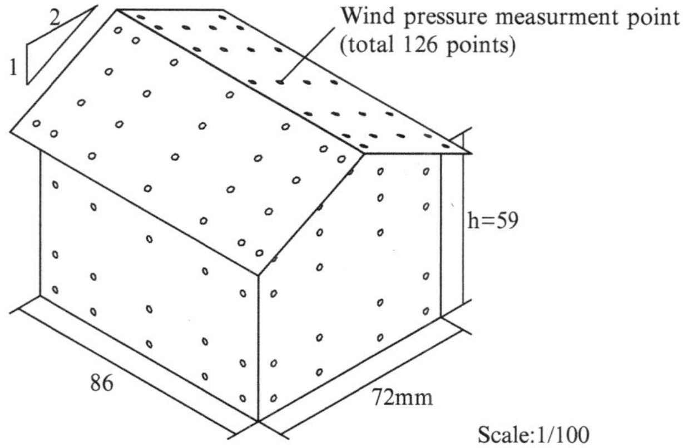

Fig. 4-2. The target model of wind tunnel experiment

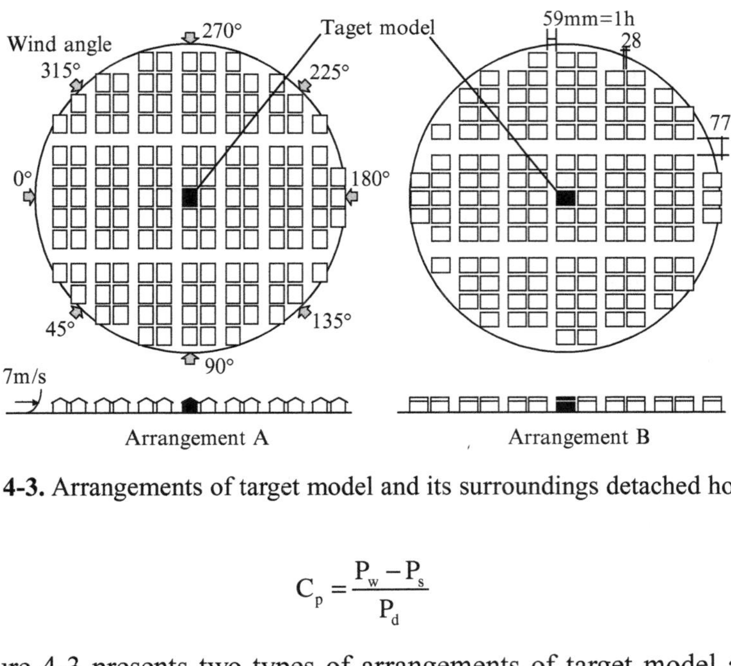

Fig. 4-3. Arrangements of target model and its surroundings detached houses

$$C_p = \frac{P_w - P_s}{P_d} \tag{4.1}$$

Figure 4-3 presents two types of arrangements of target model and its surrounding detached houses. Arrangements A and B indicates the situations that the long-side walls and the short-side walls (gable walls) of the target model is adjacent to the road, respectively. The amount of surrounding houses was determined by the maximal possible numbers of houses on the turntable in the wind tunnel. All surrounding houses shared the same characteristics as the target model except for the measurement points. The building-to-land ratio for all houses was 60%. One block was consistent of

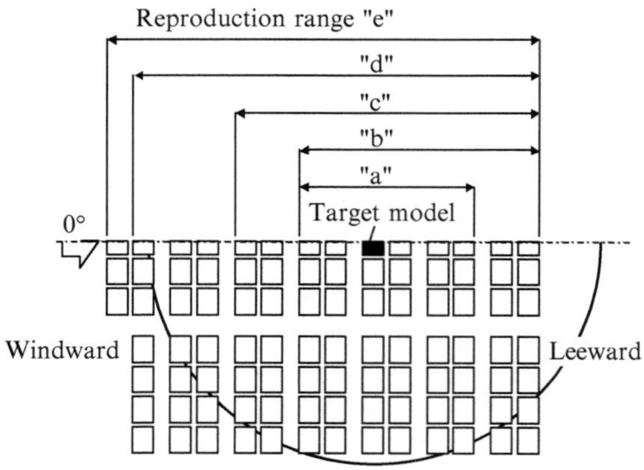

Fig. 4-4. Five patterns (i.e., a, b, c, d, and e) of reproduction ranges

10 houses (see Fig. 4-3) and the width of front road was the same as the eaves height. The road width between each block was the same as the height of the house. The diameter of the turntable is 1,600 mm, which is 27 times of the eaves height.

First, we examined reproduction ranges of the two arrangements. Fig. 4-4 shows five patterns (i.e., a, b, c, d, and e) of reproduction ranges. The pattern "a" has the smallest reproduction range; additional houses on the leeward side of the pattern "a" characterized the pattern "b." Additional houses on the windward side of the pattern "b" characterized the patterns "c," "d," and "e." The experiment measured the wind pressure coefficients from the target model in the five patterns with wind direction set as 0°.

Figure 4-5 shows the difference between the five patterns of reproduction ranges and the arrangement A. C_{pw} indicates wind pressure coefficients measured in arrangement A. ΔC_{pw} indicates the difference of wind pressure coefficients measured between the five patterns of reproduction ranges and the arrangement A. The result showed that ΔC_{pw} measured from the difference of the patterns "a" and "b," and arrangement A was significant.

In contrast, the ΔC_{pw} between the patterns "c," "d," and "e," and arrangement A were smaller. Accordingly, it is appropriate to investigate wind pressure coefficients of the target model when three blocks are added to both leeward and windward sides of it. Furthermore, in arrangement B, because ridgepoles of each house were parallel to wind flows of the wind tunnel, it is appropriate to use arrangement B as a representation of urban areas as well.

Figure 4-6 presents wind pressure coefficient distribution of the target model in arrangements A, B, and without surrounding houses (i.e., case S) in wind direction of 0°. The distribution of wind pressure coefficients on

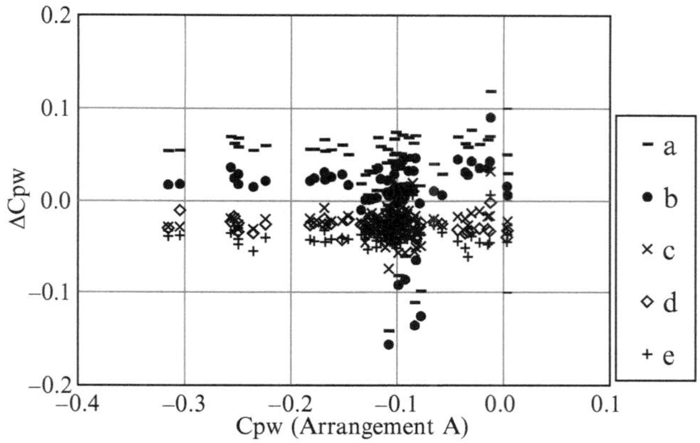

Fig. 4-5. The difference of wind pressure coefficient (ΔC_{pw}) between the five patterns of reproduction ranges and the arrangement A

the windward wall and gable walls was wildly ranging in case S. In contrast, the values of wind pressure coefficients on each wall in arrangements A and B were narrowed distributed and close to zero. The difference of average wind pressure coefficients between windward and leeward walls was 0.94 in case S and was 0.02 in arrangements A and B. The maximal difference of wind pressure coefficients between windward and leeward walls was 0.06. Accordingly, it is difficult to ensure enough cross-ventilation in houses located in the congested urban areas if only relying upon openings on the wall. On the other hand, wind pressure coefficients measured on roofs were negative regardless of wind directions or surrounding house arrangements. When comparing arrangements A and B, the absolute values and the distribution of wind pressure coefficients were slightly different. However, the absolute values of wind pressure coefficients on walls were much smaller and wind pressure coefficients on roofs were negative for all conditions.

4.1.3 Improving-Techniques of Cross-Ventilation in Detached Houses

Next, techniques for improving cross-ventilation effect by using negative pressure on roof surfaces in detached houses in arrangement A were discussed. Fig. 4-7 presents four cases with different techniques [i.e., void (1), void (2), monitor roof, and wind tower].

Figure 4-8 shows the plan and exploded Fig. of the void (1). The distribution of wind pressure coefficients on wall A is shown by heights on Fig. 4-9.

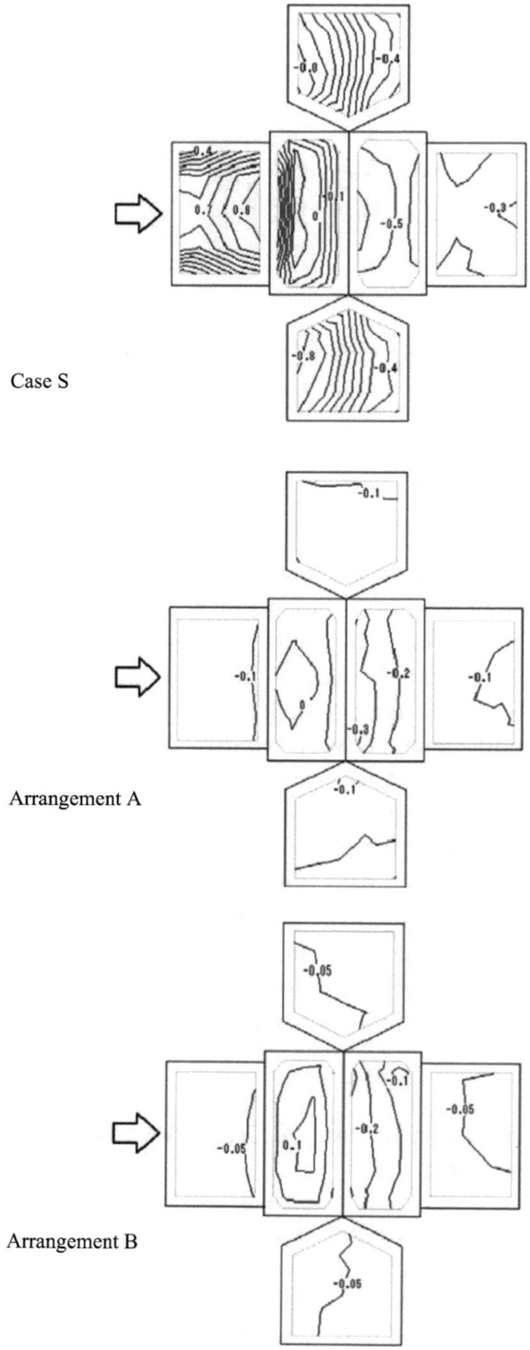

Fig. 4-6. Wind pressure coefficients distribution of the target model in arrangements A, B, and without surrounding houses (i.e., case S) in wind direction 0°

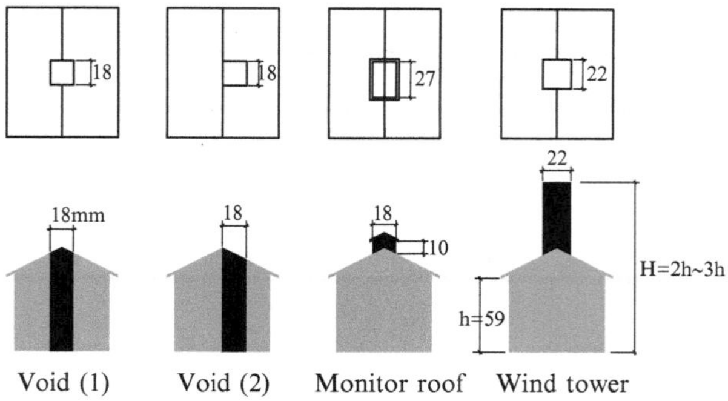

Fig. 4-7. Test cases of techniques for improving cross-ventilation

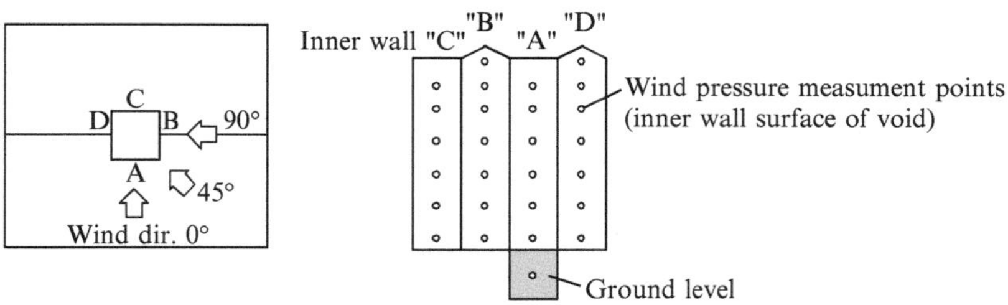

Fig. 4-8. The plan and exploded figure of the void (1)

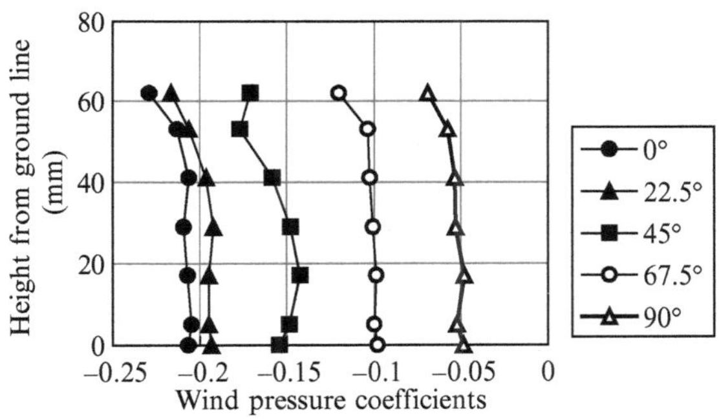

Fig. 4-9. The distribution of wind pressure coefficients on wall A

The wind pressure coefficients were similar regardless of varied heights. The average values of wind pressure coefficients of all void walls (A, B, C, and D) were similar regardless of varied wind directions (see Table 4-1). Accordingly, the distribution of wind pressure coefficients inside the void was restricted. As a result, the following discussion was based on the average of wind pressure coefficients inside the void (C_{PV}).

Table 4-1. The average values of wind pressure coefficients of all void walls

Wall	0°	22.5°	45°	67.5°	90°
A	−0.21	−0.20	−0.16	−0.10	−0.66
B	−0.21	−0.20	−0.16	−0.11	−0.66
C	−0.21	−0.20	−0.16	−0.11	−0.66
D	−0.22	−0.21	−0.16	−0.11	−0.05

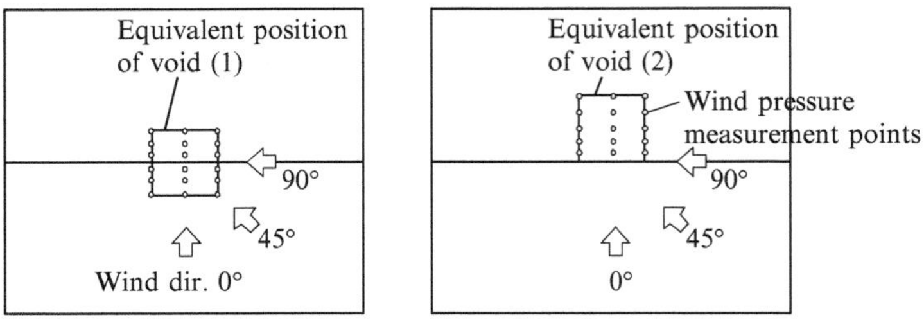

Fig. 4-10. Measurement points on the roofs of the void in case void (1) and (2)

Furthermore, we examined whether wind pressure coefficients inside the void can be obtained from wind pressure coefficients measured on the roof of the void. Fig. 4-10 shows locations of measurement points on the roofs of the void in case void (1) and (2). The wind pressure coefficients on the roof (C_{PR}) were represented by the average wind pressure coefficients measured by these measurement points. Figs. 4-11 and 4-12 present results of C_{PR} and C_{PV} in case void (1) and (2). In each case, the values of C_{PR} and C_{PV} were similar regardless of wind directions. Therefore, wind pressure coefficients inside the void are likely to be obtained from wind pressure coefficients measured on the roof of the void.

We next examined the monitor roof. The wind pressure coefficient of monitor roof (C_{PM}) was represented by the average wind pressure coefficients of the front wall of the monitor wall as shown in the grey area of the left sided model in Fig. 4-13. The wind pressure coefficient of normal roof (C_{PR}) was represented by the average wind pressure coefficients of the roof, which was shown in the grey area of the right-sided model in Fig. 4-13. Values of C_{PM} and C_{PR} in different wind directions are shown in Fig. 4-14. The more measurement points toward the leeward side, the larger negative wind pressure coefficients were obtained in both cases. C_{PM} became −0.35 when measurement points faced to the leeward side. The absolute values of C_{PM} were larger than C_{PR} regardless of wind directions. Although the wind

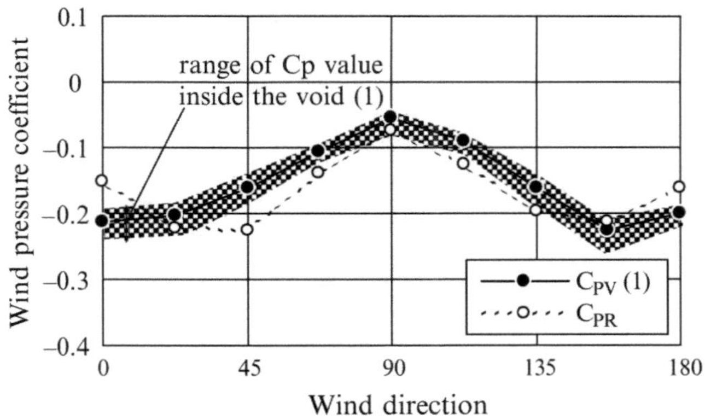

Fig. 4-11. Comparison of C_{PR} and C_{PV} in case void (1) according to the wind directions

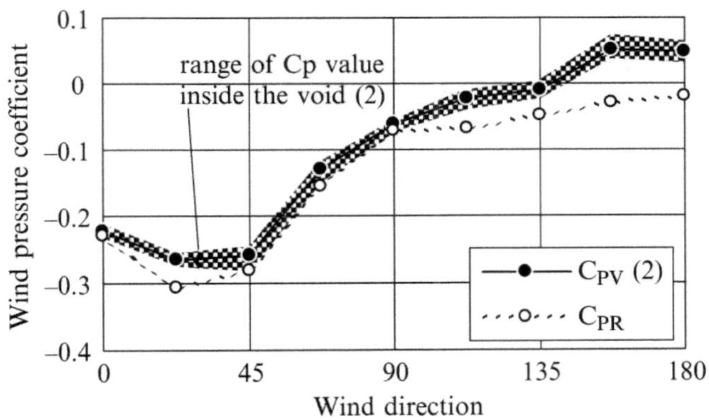

Fig. 4-12. Comparison of C_{PR} and C_{PV} in case void (2) according to the wind directions

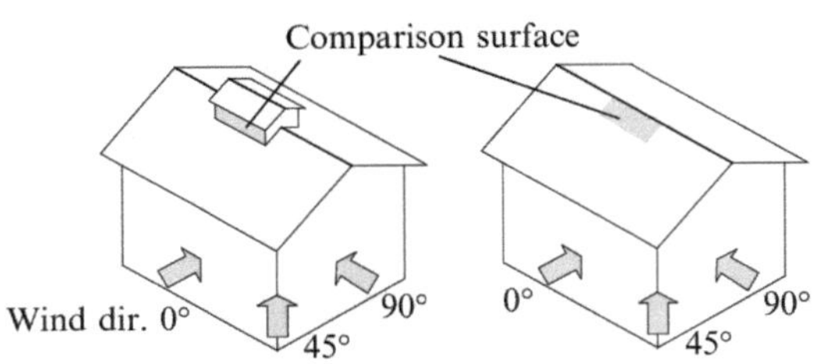

Fig. 4-13. Comparison surface of the wind pressure coefficients

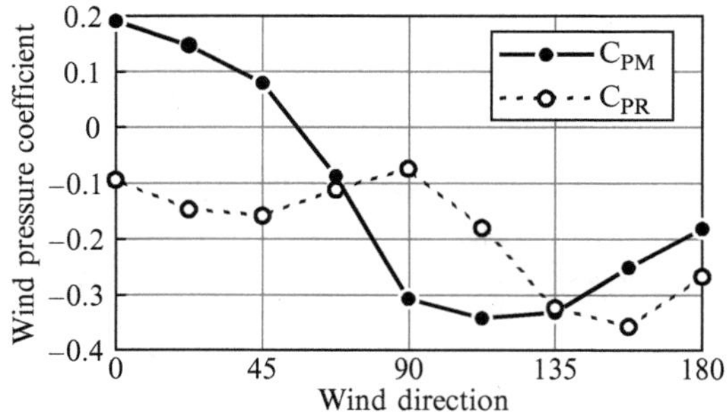

Fig. 4-14. Comparison of C_{PM} and C_{PR} according to the wind directions

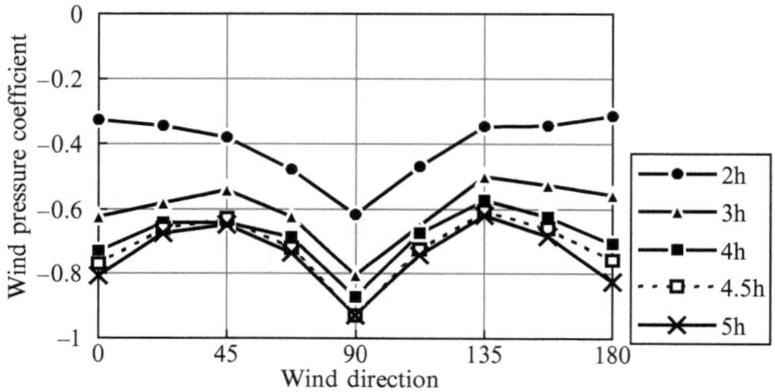

Fig. 4-15. The relationship between the heights of the wind tower and the wind pressure coefficients on the top surface according to the wind directions

pressure coefficients of normal roof were similar regardless of wind directions, the monitor roof was more efficient in improving cross-ventilation because it produced larger negative pressure.

The relationship between the heights of the wind tower and the wind pressure coefficients on the top surface of the wind tower is shown in Fig. 4-15. The wind pressure coefficients on the top surface of the wind tower were always negative. The wind pressure coefficients tended to converge toward a certain value with larger height of the wind tower. Specifically, when the height of the wind tower was approximately three times of the eaves height, the wind pressure coefficients converged.

The cross-ventilation flow rates of cases [i.e., void (1), monitor roof, and normal roof] were calculated by the flow-network model using wind pressure coefficients obtained from previous experiments. The current calculation only considered the wind ventilation. Outdoor wind velocity at the eaves height was 1 m/s. Fig. 4-16 shows the plans of calculated models.

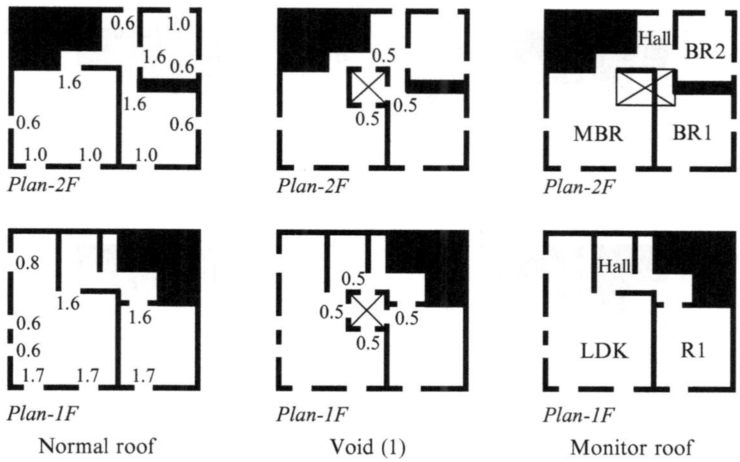

Fig. 4-16. Plans of calculated models [numbers mean opening area (m^2)]

Plans of the normal roof were slightly modified in order to be comparable to the placing of the void and monitor roof. Specifically, ridgepole was revealed in the void space and monitor roof was placed on the ceiling in the master bedroom. The hallway, corridor, and staircase were assumed to be in one room. Windows presented in Fig. 4-16 were open when doing the calculation. The size of opening faced the void was 60 cm × 90 cm; the size of opening on the wall of the monitor roof was 120 cm × 60 cm. Discharge coefficients of openings were all 0.65.

The cross-ventilation flow rates of the void case (Q_V) and the normal roof case (Q_0), which were without void and monitor roof in three different wind directions (i.e. 0°, 45°, and 90°), are shown in Fig. 4-17. In wind direction of 0° and 45°, outflows toward the void from each room change with the outflow/inflow of openings. In wind direction of 90°, outflows toward the void and outflow/inflow of openings were independent. Fig. 4-18 presents the ratio of Q_V and Q_0 in difference wind directions. Cross-ventilation improvement effects in the void case were approximately twice than those in the case of normal roof. However, in wind directions parallel to ridgepole (i.e., 90° and 270°), the difference of wind pressure coefficients between the void and exterior walls were small. Moreover, increases of cross-ventilation flow rates were limited.

As mentioned earlier, wind pressure coefficients inside the void are likely to be obtained from wind pressure coefficients on the roof of the void. Fig. 4-19 shows the cross-ventilation flow rates calculated by wind pressure coefficients of the roof and wind pressure coefficients of the void.

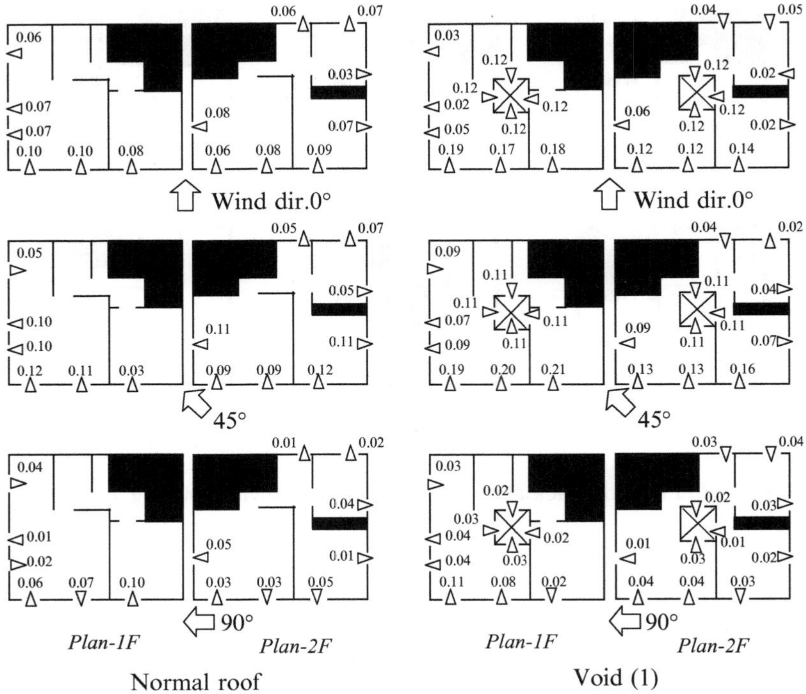

Fig. 4-17. The cross-ventilation flow rates of the normal roof case Q_0 (m^3/s) and the void case Q_V (m^3/s) in three different wind directions

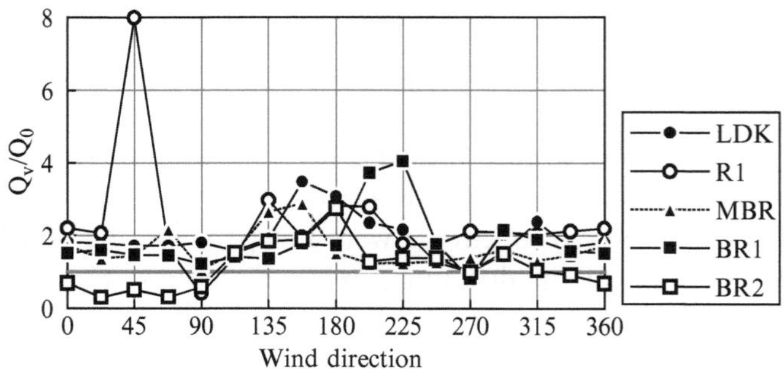

Fig. 4-18. The ratio of Q_V and Q_0 according to the wind directions

Cross-ventilation flow rates were calculated by wind pressure coefficients of the roof were slightly larger than those obtained through wind pressure coefficients of the void; however, it did not affect calculations of cross-ventilation flow rates.

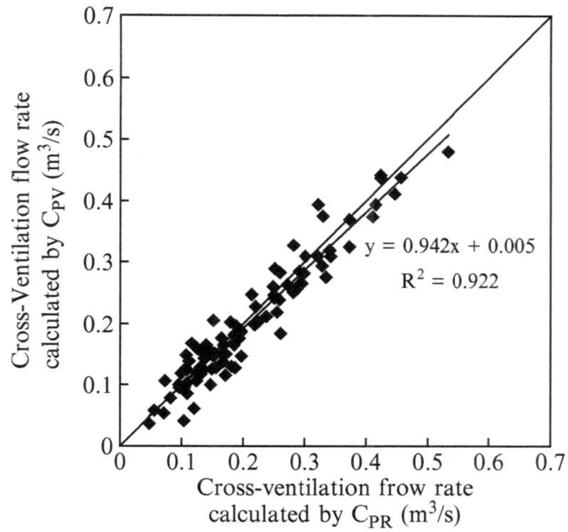

Fig. 4-19. Comparison the cross-ventilation flow rates calculated by wind pressure coefficients of the roof and wind pressure coefficients of the void

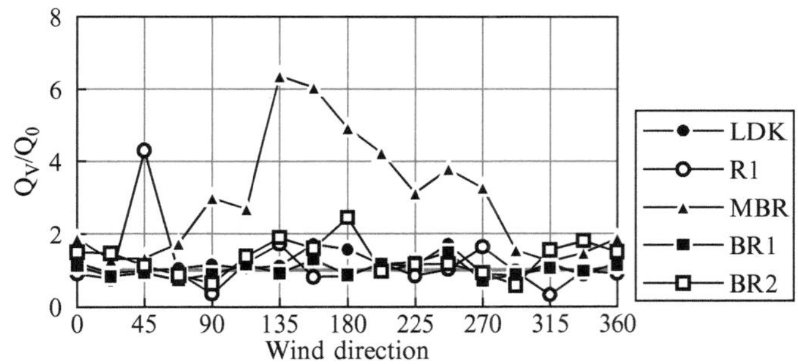

Fig. 4-20. The ratio of Q_M and Q_0 according to the wind directions

The ratio between cross-ventilation flow rates of the monitor roof case (Q_M) and Q_0 is shown in Fig. 4-20. Although the improvement effect of cross-ventilation only showed in some rooms, the maximal cross-ventilation flow rate became more than six times compared to the case of normal roof (e.g., in the master bedroom). Accordingly, cross-ventilation improvement effect due to monitor roof is very promising if with proper consideration of the locations of monitor roof in the house plan.

Figure 4-21 shows the comparison of cross-ventilation flow rate of each room averaged at wind-angle between building-cases. In the case of void (1), the flow rate increases compared with the case of normal-roof in all rooms, and improving cross-ventilation effect is remarkable. In the case

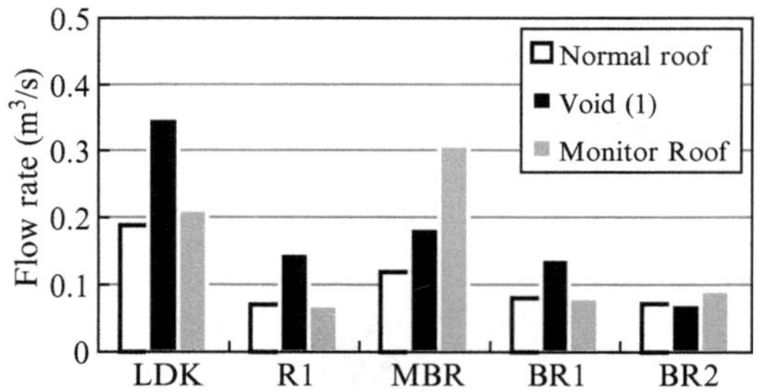

Fig. 4-21. The ratio of Q_M and Q_0 according to the wind directions

of monitor-roof, improving cross-ventilation effect is remarkable only in the MBR, which exists under the monitor roof. As mentioned above, void is very effective improving-techniques of cross-ventilation in detached houses, if there is much space in the building.

4.2 Cross-Ventilation Utilization of the Housing in Congested Urban Area in Taiwan

4.2.1 Introduction

Taiwan has the second largest population density among countries or regions with 10 million people or more in the world. The population density tends to be steady due to decreased birthrate over the past several years. However, the population keeps concentrating in the cities rather than rural areas due to the convenience and the modernization. Due to the overcrowded population, increased energy consumption causes the thermal environment problems, including the heat island phenomenon. Coolers are used frequently in Taiwan due to its high temperatures and humidity. This has contributed to increased energy consumption and become a big concern. In fact, the possession rate of coolers in the country has increased from 52% in 1991 to 85% in 2003 and the rate is more than 90% in the urban areas. Moreover, the energy consumption from cooler utilization in general housing is increasing every year. Accordingly, it becomes crucial to develop techniques that can diminish excessive energy consumption. Based on the study conducted by Chou (Chou 2000), the period for residents to be able to use cross-ventilation is quite long in Taiwan. It will be reasonable to expect to use cross-ventilation as an alternate and effective way to overcome the disadvantage brought by cooler utilization. Indeed, some studies have investigated utilization of cross-ventilation in Taiwan; however, they tended to focus on middle or high rise buildings (Chen et al. 2001). Moreover, some studies did attempt to investigate the utilization of cross-ventilation in the low rise house, which is more common in the country, by using only single house models (e.g., Chou 1995). To date, there is no study investigating issues associated with the utilization of cross-ventilation in low rise houses that factors in the surrounding congested buildings in the area. Thus, we aim to investigate issues related to the utilization of cross-ventilation in terraced houses, which are the most common style of low rise houses in Taiwan's urban areas.

4.2.2 Weather Conditions in Taiwan

This study examines using the cross-ventilation utilization in both a ventilation capable period and non-ventilation capable period based on the meteorological data collected between 1992 and 2001. Our data shows that the velocity of the wind during the non-ventilation capable period was slower than the ventilation capable period that lasts for 4 or 5 months/year. The average wind velocity during the ventilation capable period of the whole country was 2.1 m/s. We then divided 14 main cities in Taiwan into four groups based on their average cumulative distribution frequency (i.e., CDF) of the wind velocity of all the cities (Fig. 4-22 and Table 4-2).

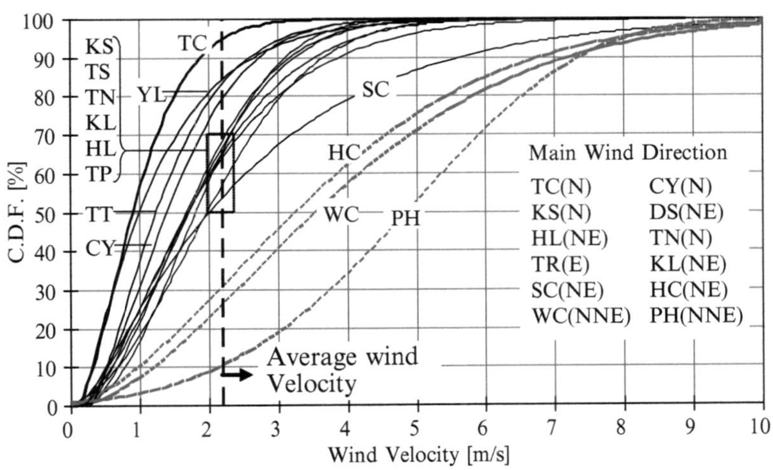

Fig. 4-22. Cumulative distribution frequency of the wind velocity during ventilation capable period

Table 4-2. Characteristics of cities based on CDF of the wind velocity

Group	CDF of the wind velocity at 2.1 m/s	City name	Wind velocity
I	>90%	Taichung (TC)	Very low
II	70–90%	Yilan (YL), Taitung (TT), Chiayi (CY)	Low
III	50–70%	Kaohsiung (KS), Tamsui (TS), Tainan (TN)Keelung (KL), Hualien (HL), Taipei (TP) Hsinchu (SC)	Medium
IV	<50%	Hengchun (HC), Wuchi (WC), Penghu (PH)	High

4.2.3 Characteristic of Houses in Taiwan and the Questionnaire Survey on Those Residents' Consciousness of Cross-Ventilation

It is crucial to use questionnaire surveys because it serves as an effective way for the researchers to gather information about local residents' concerns, lifestyles, and unit plan preferences. These factors affect residents' utilization of cross-ventilation.

The questionnaire survey has three specific aims. First, we aim to understand the opening characteristics in Taiwan. Second, we want to gather information regarding ventilation utilization during late autumn (i.e., October and November). It is predicted that the rate of ventilation use should be higher during the ventilation capable period in contrast to the non-ventilation capable period based on our previous analysis of the meteorological data. Finally, we want to investigate the air-conditioning usage behaviors, including when residents begin and stop using it (i.e., ending time) as well as the preset degrees of the air-conditioning systems in residents' living rooms or bedrooms during summer time.

4.2.3.1 Outline of Investigation

In cooperation with other researchers, the data from the questionnaire survey was collected mainly from college students in Taiwan. The outline of the investigation can be found in Table 4-3.

4.2.3.2 Characteristics of Opening and Shutting the Openings

The results suggest that over 90% of residents have double sliding windows or terrace doors. Over 80% of the residents have window screens regardless of their regions or the house styles as shown in Figs. 4-23 and 4-24. Thirty-five

Table 4-3. The outline of the investigation

Investigation period (2003/10/18 to 2003/12/17)	**Taipei area**	**Taichung area**	**Kaohsiung area**	**Total**
Questionnaires sent out	120	120	300	540
Collections	95	93	252	440
Effective collections	84	90	213	387
Housing complex	80	31	124	235
Terraced housing	3	46	87	136
Single house	1	13	2	16

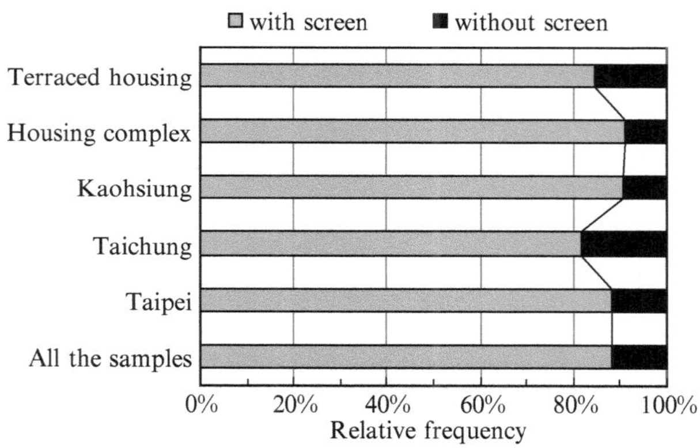

Fig. 4-23. The rate of using screens in various regions and housing

percent of the housing complexes are with grilles in windows and 48% of terraced houses have similar equipment. In addition, 10% of residents closed windows and 38% of residents opened windows all day long when being asked "Did you open or close windows during this summer?" Furthermore, the rate of opening windows when they are home becomes 65% or higher. Fig. 4-25 demonstrates the differences between the residents' responses and the data analyzed from the meteorological information in Taipei. It describes which months of a year the residents feel they can use the cross-ventilation (i.e., ventilation capable period). This suggests residents' subjective feelings of temperature and humidity differ from the meteorological data. The relationship of questionnaire results and the meteorological data are very similar from May to October. The possible explanation for the discrepancy during winter is that people in Taiwan do not typically use heater systems in the winter season.

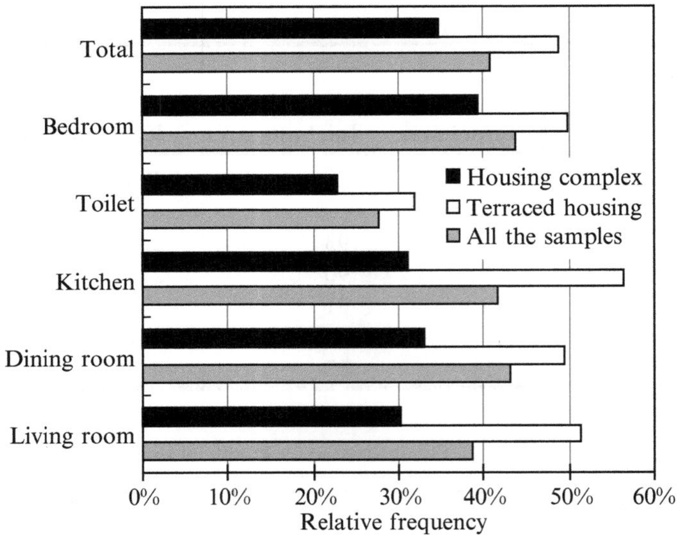

Fig. 4-24. The rate of using grilles in various rooms

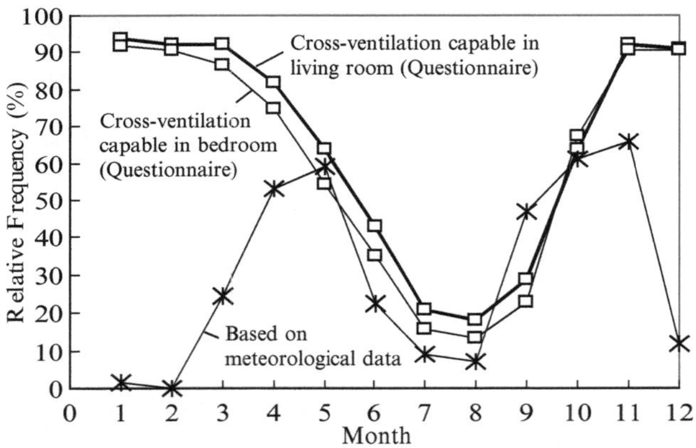

Fig. 4-25. The differences between the result of questionnaire and the meteorological data (Taipei)

4.2.3.3 Features of Air Conditioning System Using Behaviors

The results indicate 64% of participants used coolers and 66% of participants used fans during the summer. The average duration of cooler usage was approximately 110 days in 2003, and the duration became longer when moving from north to south. In Taipei (a northern city), residents mainly began using their coolers in June whereas in Taichung (a city in mid-Taiwan) and Kaohsiung (a southern city), it was in July. Taipei and

Taichung residents stopped using them in September; Kaohsiung residents in October. In addition, the results show the average preset temperature in the living room was 25.0°C, and the bedroom was 25.3°C.

4.2.3.4 Residents' Satisfaction of the Indoor Air Environment

Approximately 70% of residents feel dissatisfied to the air environment in contrast to the environment as a whole. Moreover, people who live on the fifth floor or below complained more because they felt the indoor wind velocity was low. Fig. 4-26 indicates that the problem becomes more remarkable when the building was low and should be investigated further via the wind tunnel experiments.

4.2.4 Selection of the Type of Houses for Further Investigations

Figure 4-27 shows the congested urban area in Kaohsiung and an example of the terraced house. In order to avoid rainfall, the front side of the first floor of the terraced house is designed approximately 3 m away from the road which creates a space called "Chilo," (currently restricted by Building Standard Law), and the height of first floor is always higher than the other floors. In addition, one house always connects with the other house(s), and

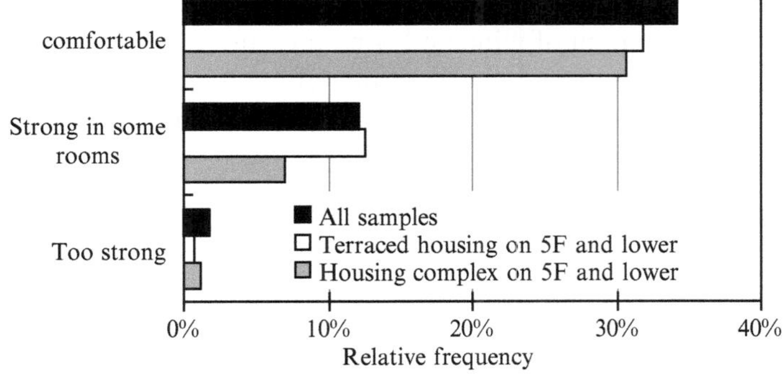

Fig. 4-26. Residents' satisfaction of the indoor air velocity

Fig. 4-27. The congested urban (*above*), the outfit of the terraced house (*left*) and "Chilo" (*right*)

the openings are set up only in the back and front sides. Stairs are usually located in the center of the house. A three-story terraced house was used as the model in the current study.

All the wind tunnel experiments were conducted on the terraced house models in the congested urban areas of Taipei and Kaohsiung. We measured the effects of the amount of wind and wind pressure coefficients on ventilation. Sub-sequentially, based on the results of the wind tunnel experiments, the amount of cross-ventilation is calculated by an airflow network model.

4.2.5 Wind Pressure Coefficient of the Target Buildings Measured on the Wind Tunnel Experiment

4.2.5.1 Experiment Settings

The scale in our experiments was 1/80, which was the smallest possible scale. There were a total of 126 measurement points to record the wind pressure coefficients on the surfaces of the model. We manipulated four cases of arrangements (i.e., Single, Raw, Block, and Area) based on the number of buildings surrounding the test building. The "Area" case had the highest resemblance to the current congested urban living area (Fig. 4-28). All four cases and the types of openings were tested using 16 different wind directions to examine the wind pressure coefficients. Then we calculated the flow rate based on the room pressures, which were obtained via the Newton-Raphson method.

4.2.5.2 Distribution of Wind Pressure Coefficient (Cp)

The pressure values were indicated by the different colors as shown in Fig. 4-29. Positive pressures were indicated by "hot" colors (e.g., red, orange) and the negative pressures were indicated by "cold" colors (e.g., blue, green). In the "Single" arrangement condition with a wind direction of 0°, the results showed that the maximal Cp of windward side was 0.8 and the Cp on the wall side of the Chilo was positive. Moreover, the minimal Cp of leeward side was −0.3 and the Cps of flat roof were ranging from −0.6 to 0.3. When the wind direction changed to 45°, the results demonstrated that the maximal Cp of windward side was 0.3 and the Cp on the wall side

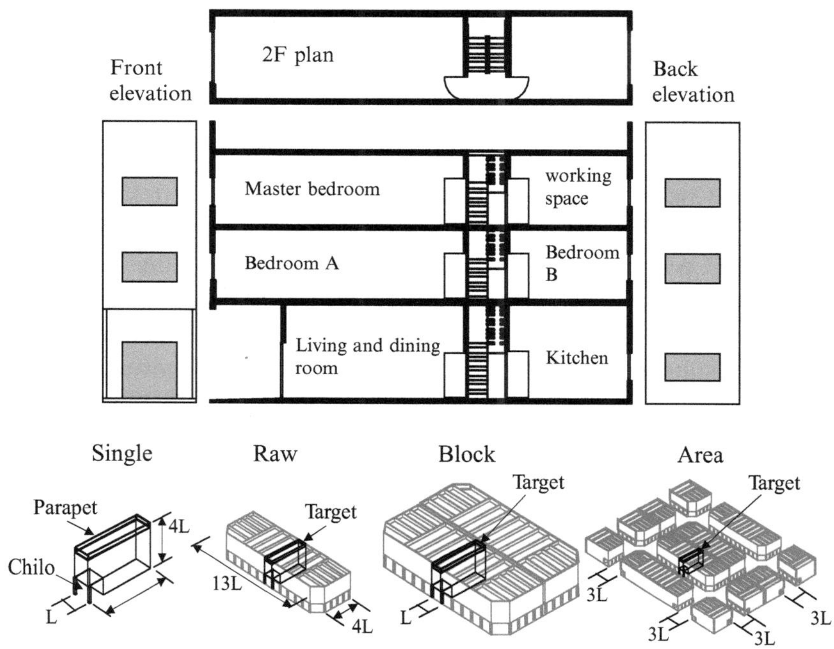

Fig. 4-28. Plan, section and elevations of target model and arrangements for wind tunnel experiments

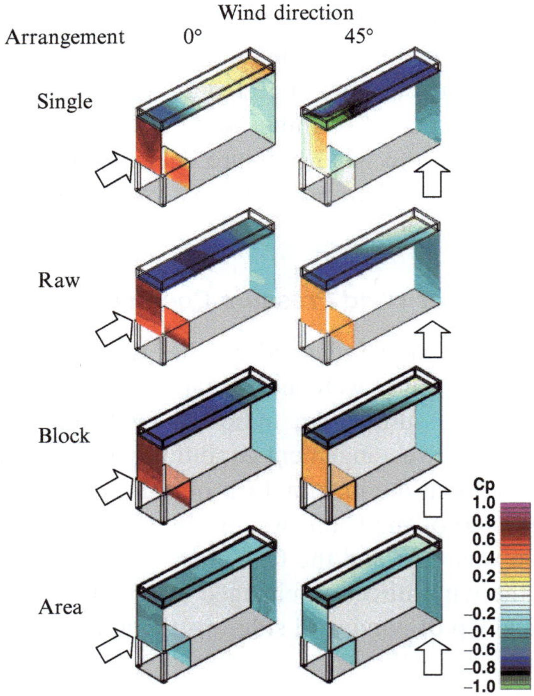

Fig. 4-29. Arrangements for wind tunnel experiments

of the Chilo was negative. Moreover, the minimal Cp of leeward side was −0.6 and the Cps of flat roof were ranging from −0.6 to −1.0. In the "Raw" arrangement condition with a wind direction of 0°, the results showed that the maximal Cp of windward side was 0.8 and the Cp on the wall side of the Chilo was positive. Moreover, the minimal Cp of leeward side was −0.4 and the Cps of flat roof were negative. When the wind direction changed to 45°, the results demonstrated that the maximal Cp of windward side was 0.3 and the Cp on the wall side of the Chilo was also 0.3. Moreover, the minimal Cp of leeward side was −0.6 and the Cps of flat roof were ranging from −0.1 to −0.8. The results found in the "Block" arrangement condition showed the Cps of windward, leeward, and roof are all negative when the wind direction was 180°. In the "Area" arrangement condition, the Cps of windward, leeward, and roof all showed negative values in the 16 wind directions. The absolute values of the Cp on the wall side of the Chilo were smaller than those on the wall of windward or leeward side. The difference of Cp between windward and leeward sides was less than 0.1.

4.2.6 Prediction of Cross-Ventilation Flow Rate in the Target Buildings

Cross-ventilation flow rates were calculated from the Cps obtained from the experiments by using the Newton-Raphson method. Parameters relevant to the calculation were listed in Table 4-4. The discharge coefficients of all openings of the model were set as 0.67. Each story was considered as one unit. The cross-ventilation flow rate indicated the sum of flow rates of the three stories and represented the overall flow rate of the whole building. Fig. 4-30 shows the results for each of the arrangements (i.e., Single, Raw, Block, and Area). The results demonstrated that the "Area" arrangement,

Table 4-4. Calculation settings

Openings	Real opening area
Entrance	1.73 m^2
Other windows	0.86 m^2
Staircase area	3.20 m^2
Lantern	2.88 m^2
Area of top of wind tower	2.88 m^2
Interior doors	2.00 m^2
Standard height	12 m
Standard wind velocity in Taipei	2.34 m/s
Discharge coefficient	0.67

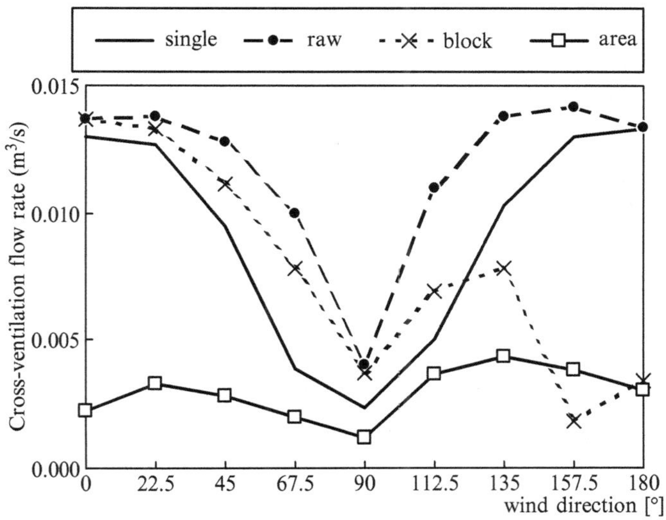

Fig. 4-30. Cross-ventilation flow rates of all cases

which was closest to the real urban areas when compared to the other three arrangements, generated the smallest flow rate, particularly when the wind directions were 0° and 180°. Furthermore, the velocity of the openings became 0.1 m/s on average if adding the partition into calculation. So the velocity is not significant enough to have an impact on the utilization of the ventilation. Overall, our findings from the wind tunnel experiments had similar results as obtained from the survey, both suggesting the indoor wind velocity was low. Based on the results described above, the improvement of cross-ventilation was limited by changing the openings, and the roof would gain greater wind pressure. Results also indicated that changing the roof designs could increase the cross-ventilation flow rate.

4.2.7 Strategy for Improving Cross-Ventilation in the Target Buildings

4.2.7.1 Experiment Settings

In order to examine the relationship between the roof design and the cross-ventilation flow rate, three roof designs were used: (1) lantern, (2) gable roof with a lantern, and (3) wind tower. Specifically, the lantern in design 1 was set at the flat roof directly above the staircase (Case-PL). In design 2, the additional gable roof was added to the top of the model with a lantern located just above the staircase (Case-RL). In design 3, varying of heights with rectangular pillar shaped wind towers were added directly above the staircase (Case-V1 ~ 7).[1] Furthermore, we focused on examining the variables described above in the Area arrangement. Similar to previous experiments, Cps were obtained from the wind tunnel experiments and the average cross-ventilation flow rates of all directions were calculated from Cps by using the Newton-Raphson method.

4.2.7.2 Experiment Results

The results demonstrated that with the lantern design, the flow rate was 2.36 times higher than with the original design. Using the gable roof with a lantern condition, the flow rate was 3.94 times higher than in the original

[1]Heights of the wind tower of Case-V1~7 were H, 4/3H, 5/3H, 2H, 7/3H, 8/3H, and 3H, respectively. H indicated the height of one floor, 360 cm.

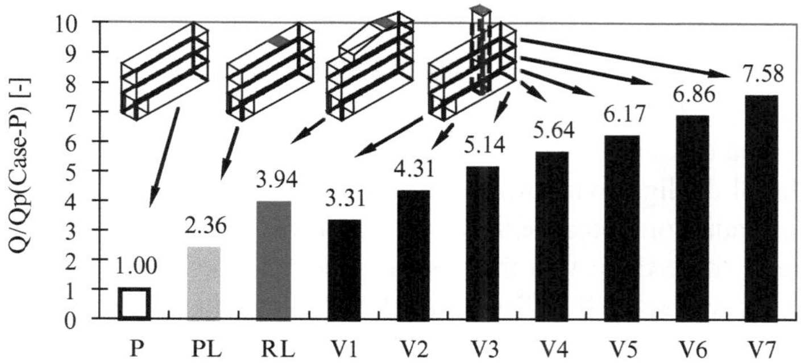

Fig. 4-31. Comparison of cross-ventilation flow rate of all cases

condition. Finally, in the wind tower condition, the flow rate was 3.31~7.58 times higher than in the original condition, depending upon the heights of the towers (Fig. 4-31). In one case (Case-V4) of the wind tower condition, the Cps changed with the wind directions on the side walls of the wind tower. However, the Cps of the top of the wind tower remained stable regardless the wind directions. Thus, it is necessary to consider the direction of prevailing winds if the opening is built on the side wall of the wind tower. Theoretically the higher the wind tower, the greater the flow rate. In reality, however, a wind tower less than two-stories high in a three-story building seems to be reasonable.

4.2.8 The Effect on Energy Conservation Due to the Improvement of the Cross-Ventilation

The numerical simulation of the thermal and airflow network model, AE-Sim/Heat, was used to calculate the effect of energy conservation through improvement of the cross-ventilation. Three cases of terraced house models were used for the simulation. The first case (Case P) was the original condition without cross-ventilation; the second case (Case PV) was the original condition with cross-ventilation, and the third case (Case-V4) was the original condition with a two-story wind tower and cross-ventilation.

Standard weather data was based on the information obtained in Taipei city. The wind pressure coefficients were based on the results obtained in the Area arrangement experiment. The cooling loads (heat removal) of the model were calculated via the numerical simulation in order to evaluate the effect on the improvement of the cross-ventilation.

4.2.8.1 Calculation Settings

The opening area was 0 m^2 for Case P, 4.32 m^2 for the Case PV, and 7.20 m^2 for the Case-Vsh090. The daylight areas of the three cases were 20.16 m^2. The radiation area of Case P and Case PV was 97.2 m^2, and was 160.6 m^2 for the Case-V4. All daylight windows were assumed to receive both direct and diffused solar radiation. Because the assumption was established in the congested urban area, where there was more solar shading. The shading coefficient of the windows was set as 0.2. The schedules of occupant numbers, lighting heat generation, and machine heat generation are presented in Fig. 4-32. Annual outdoor temperature and humidity data of Taipei is also shown in Fig. 4-33. The minimal ventilation airflow rate of all rooms was set as 0.5 times/h. Kitchen fans and curtain utilization were not taken into account. The discharge coefficient of all openings was set as 0.67. It was assumed that there was no infiltration around the daylight model. The width of the slit of the entrance was set as 2 mm and its discharge coefficient was also 0.67.

Finally, the air conditioning of all cases was a hybrid ventilation system. The setting temperature of the cooler was set at 27°C. If the room temperature was above 27°C, the cooler would be operated and the windows would be closed. Cooling loads (heat removal) were simply calculated at a room temperature set at 27°C with a relative humidity of 60%.

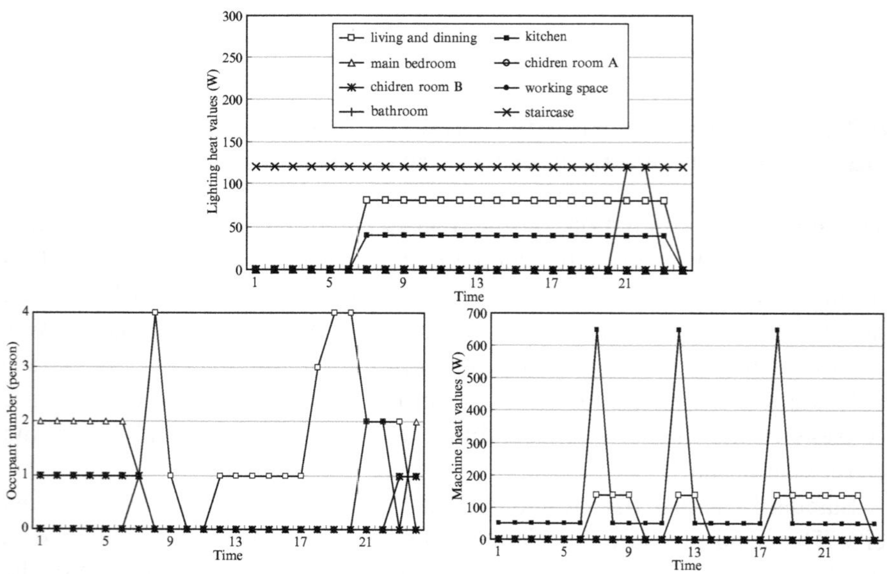

Fig. 4-32. The schedules of occupant numbers, lighting heat generation, and machine heat generation

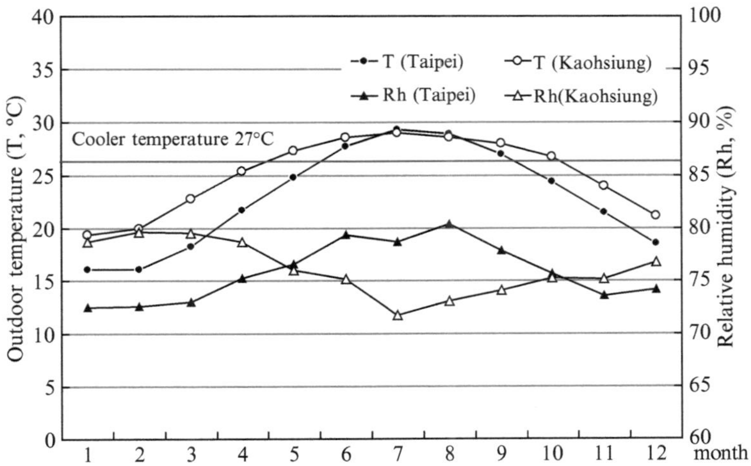

Fig. 4-33. Annual outdoor temperature and humidity data of Taipei and Kaohsiung

4.2.8.2 Calculation Results

The monthly accumulated values of sensible and latent heat removal in the three cases by the cooler are presented in Fig. 4-34. Because the cross-ventilation was unable to be operated in July and August, the heat removal of the three cases did not differ significantly. There were sensible reductions of heat removal by the wind tower cross-ventilation in May, June, September, and October, as well as significant latent heat removal in May and August. In May and October, the values of sensible heat removal were close to 0. The annually accumulated values of heat removal (sensible and latent) of the three cases are shown in Fig. 4-35. Compared to Case P, the heat removal of Case V4 decreased 16%. The heat removal of Case V4 was 39% smaller than Case P without considering the data of July and August.

4.2.9 Conclusions

The results of based on the questionnaire survey indicate that it is very common for the residents in Taiwan to have the screen and the grille in the open position which may affect the cross-ventilation. Therefore, it is essential and necessary to consider the effect of the existence of the screen and the grille in these openings when conducting a wind tunnel experiment and calculating the results. The data also suggests that the Taiwanese open their windows not only during the ventilation capable period but also during the non-ventilation capable period, such as summer. The effect of natural ventilation caused by the ventilation driving force is extremely small given the layout of the

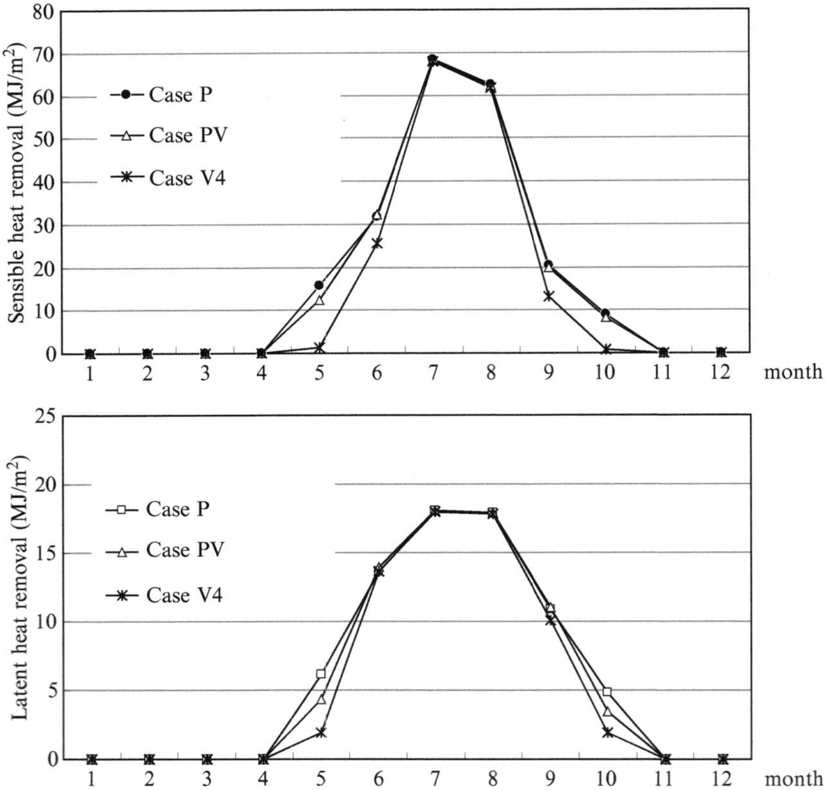

Fig. 4-34. Monthly accumulated values of sensible and latent heat

present type of buildings. The best way to improve the cross-ventilation in such houses is to install ventilation devices on the roof, because such changes can produce greater wind pressure and therefore improve the ventilation. If a two-story wind tower is added on top of the roof, the cross-ventilation flow rate can increase more than five times. Plus, compared to current conditions, it can decrease the heat removal by 16% annually.

4.3 Utility Cross-Ventilation in Guangzhou and Shenzhen in China

In this section, we have focused on the feasibility of cross-ventilation due to natural ventilation potential, specifically in the regions of Guangzhou and Shenzhen on the Pacific coast of China, well known for being densely populated and highly energy consumed.

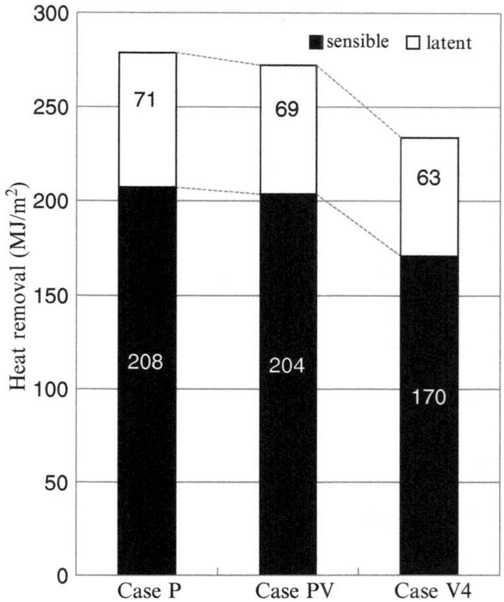

Fig. 4-35. The annually accumulated values of heat removal

4.3.1 Weather Conditions in Guangzhou and Shenzhen

4.3.1.1 Temperature and Humidity

Guangzhou and Shenzhen areas considered to have hot and humid climates. Annual mean temperature is approximately 22.2°C, while relative humidity is about 70%. In Fig. 4-36, the climograph illustrates monthly mean temperature and relative humidity in Guangzhou, Tokyo and Taipei. It can be seen that even the annual lowest dry bulb temperature and relative humidity in January are approximately 14°C and 70%, respectively. In addition, Guangzhou possesses higher annual mean temperature and humidity comparing to Tokyo as well as higher annual change in humidity level than Taipei.

4.3.1.2 Wind Velocity and Direction

In Fig. 4-37, monthly average wind velocities are illustrated. In April, wind velocity is found to reach the annual lowest level because it is in the middle of raining season. Additionally, it reaches the highest level in July during monsoon season. Annual mean wind velocity in either Guangzhou or Shenzhen is found to be around 1.5 m/s. The wind roses during summertime and wintertime in Guangzhou and Shenzhen are shown in Fig. 4-38.

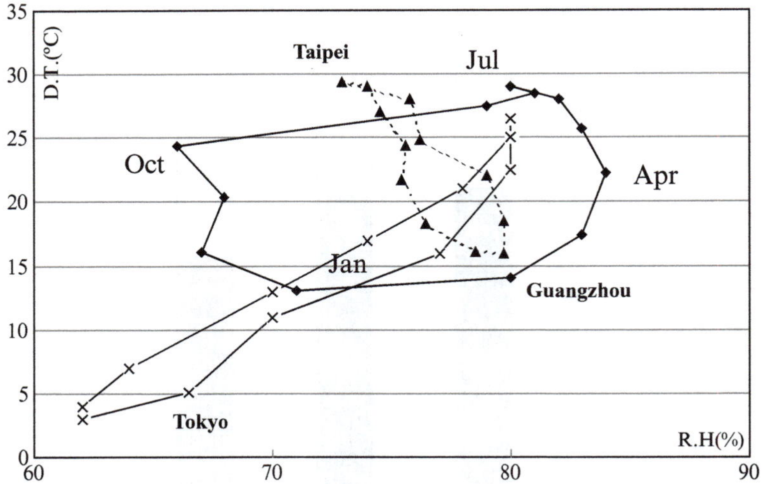

Fig. 4-36. Climograph for Guangzhou, Tokyo and Taipei

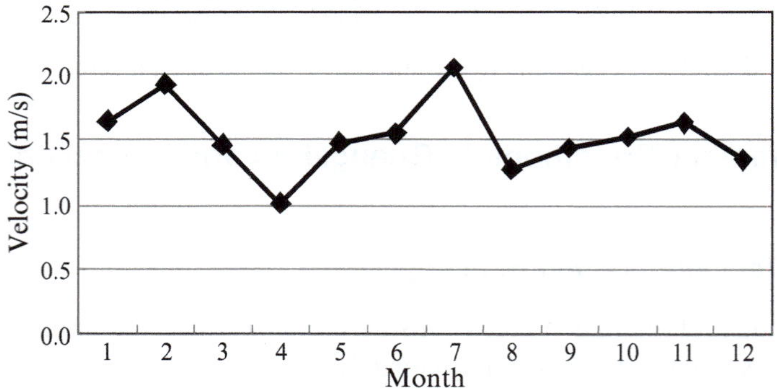

Fig. 4-37 Monthly mean wind velocity in Guangzhou

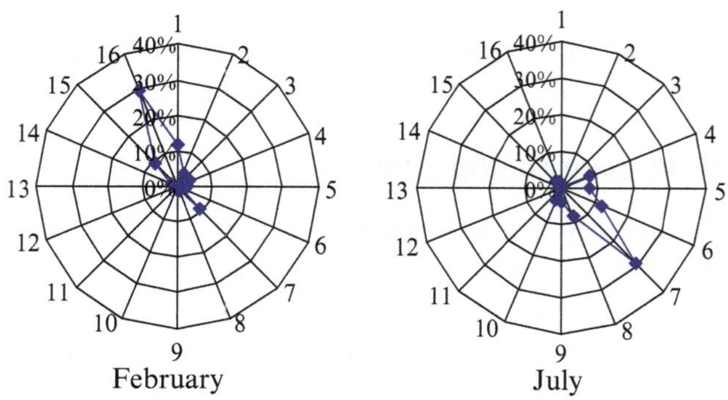

Fig. 4-38. Wind roses in Guangzhou

During summertime, prevailing wind direction approaches from southeast while during wintertime, it approaches from northwest direction.

4.3.1.3 Feasibility of Cross-Ventilation According to the Weather Conditions

The calculation of feasibility of cross-ventilation according to weather conditions was based on the standard weather data during years 1988–1997 (Zhang and Asano 2001). In addition, the conditions for calculating SET* are briefly summarized in Table 4-5. Besides temperature and humidity obtained from the standard weather data, air velocity with stagnant flow was maintained at 0.13 m/s. Clothing level altered between 0.4 and 0.9 during summertime and wintertime respectively. Reports indicate that it is feasible to utilize natural ventilation for cross-ventilation while SET* is in 22.2–25.6°C range (Gagge 1971). The frequency of occurrence of SET* felt in the ranges was calculated and illustrated in Fig. 4-39. As can be seen, during

Table 4-5. Conditions for calculating SET*

Temperature	Standard weather data in Guangzhou
Humidity	Standard weather data in Guangzhou
MRT	Identical to temperature
Air velocity	Constant at 0.13 m/s
Clothing level	0.4 clo (May–Oct), 0.9 clo (Nov–April)
Metabolic rate	1.17 met

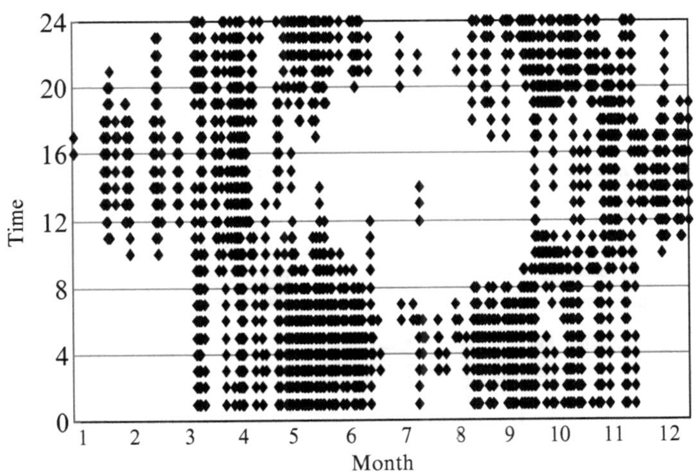

Fig. 4-39. Frequency of time in Guangzhou when SET* falls in range between 22.2 and 25.6°C where cross ventilation is considered feasible

wintertime, it can be analyzed that cross-ventilation seems to be feasible during daytime. On the other hand, it is difficult to utilize natural ventilation during summertime. In addition, there is high feasibility of implementing natural ventilation during the periods in between. In total, there are approximately 1,850 h throughout the year where feasible cross ventilation could potentially achieve comfort. Since the distance between Shenzhen and Guangzhou is about 165 km and the climate in these cities are almost same, the calculation result can be generally applied to Shenzhen.

4.3.2 Characteristic of Residential Houses in China and Questionnaire Survey on Those Residents' Awareness of Cross-Ventilation

4.3.2.1 Characteristic of Residential Houses in China

Multiple dwelling houses in China can be categorized into Tower, Corridor and Unit types which are described in detail as follows:

1. Corridor type: Equivalent to side and central corridor types in Japan (Fig. 4-40).
2. Tower type: Similar to center-core mid-to-high-rise buildings in Japan (Fig. 4-41).
3. Unit type: Close to walk up type. However it is different from cases in Japan that in this case, one staircase connects up to 3–6 units (Fig. 4-42).

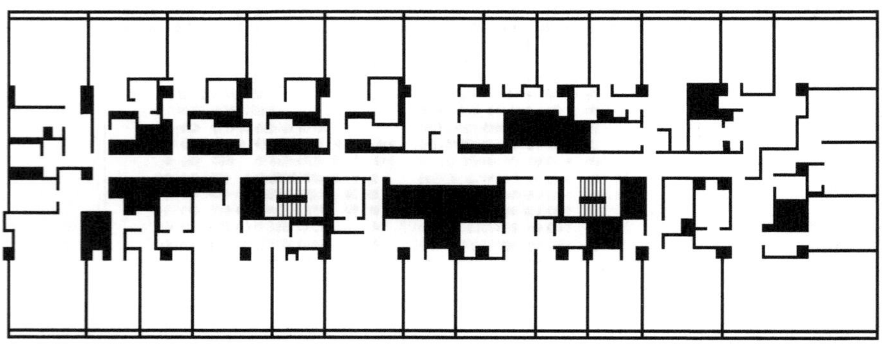

Fig. 4-40. Corridor type

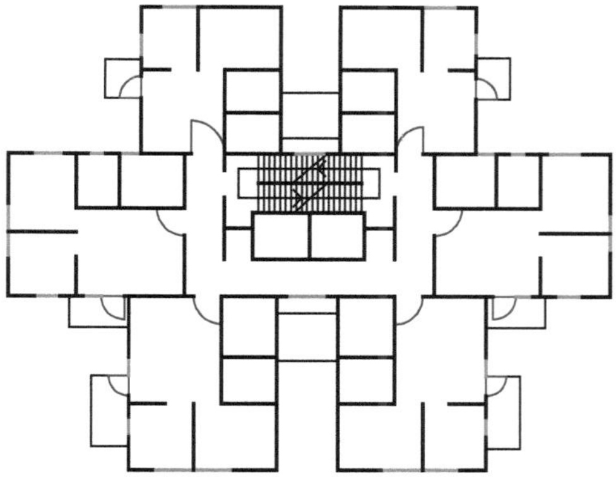

Fig. 4-41. Tower type

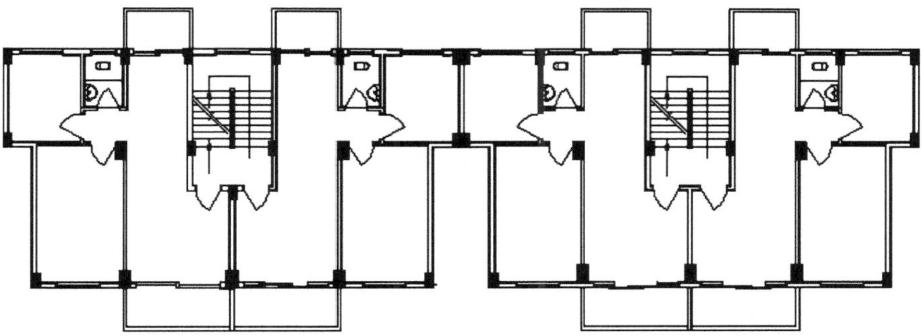

Fig. 4-42. Unit type

4.3.2.2 Questionnaire Survey on Residents' Consciousness of Cross-Ventilation

4.3.2.2.1 Questionnaire Survey Methodology

The survey was conducted from the beginning of October to December 2005 in Guangzhou and Shenzhen. The main data collected was residence types, opening configurations and conditions, use patterns of air-conditioning systems and subjects' awareness on natural ventilation. Field sampling took place with 300 questionnaires distributed in a random manner and 281 responses were obtained. The response rate was as high as 94%.

4.3.2.2.2 Results of Questionnaires Survey

Residences As illustrated in Fig. 4-43, there was high percentage of subjects in multiple dwelling houses. Within multiple dwelling houses, Tower, Corridor and Unit types possessed share of 47, 88, and 72% respectively.

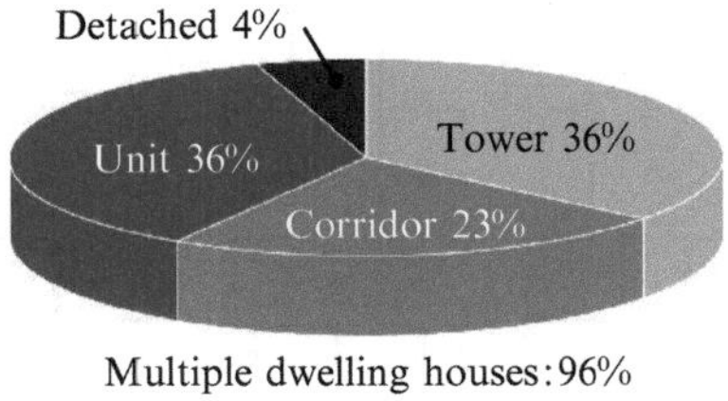

Fig. 4-43. Percentage of subjected residences

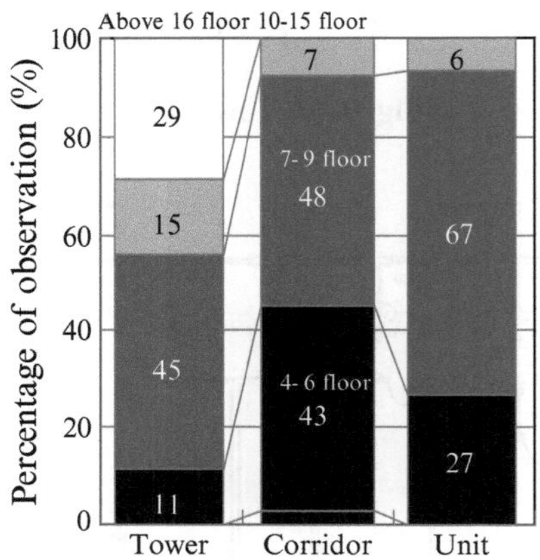

Fig. 4-44. Percentage of residences categorized by type and height

In addition, as summarized in Fig. 4-44, the majority of subjected residences in Corridor and Unit types were below ninth floor.

Opening Configurations As illustrated in Fig. 4-45, the percentage of openings with security grilles was found to be 70% in living rooms and bedrooms, while kitchens and bathrooms possessed 55%. In addition, as shown in Fig. 4-46, the residences on upper floors trend to have lower percentage of installed security grilles. Since the percentage of net window usage was about 83% in the survey in Taiwan (Tu et al. 2004), our survey at 15% was found to be comparatively low. The results led us to infer that the concern about security and pest infestation matters differs between China and Taiwan.

Opening Conditions As shown in Fig. 4-47, the percentages of either leaving windows open for the whole day or opening the windows in the presence of occupants were as high as 65% in living rooms and 56% in bedrooms.

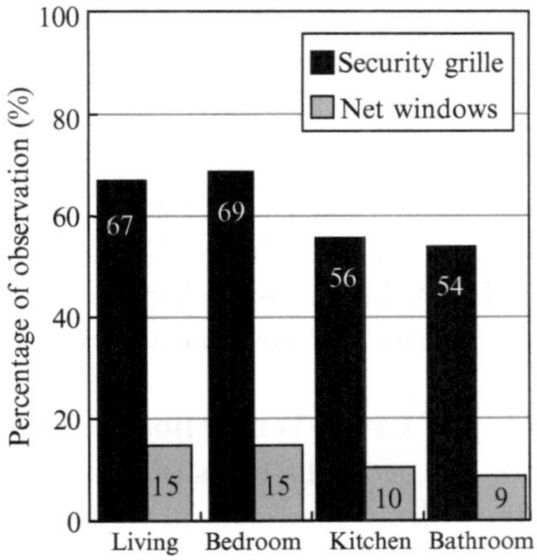

Fig. 4-45. Percentage of residences categorized by security grille and net windows

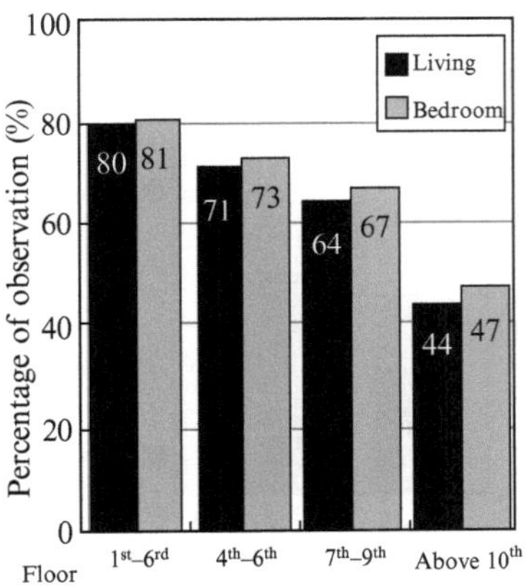

Fig. 4-46. Percentage of residence using security grille

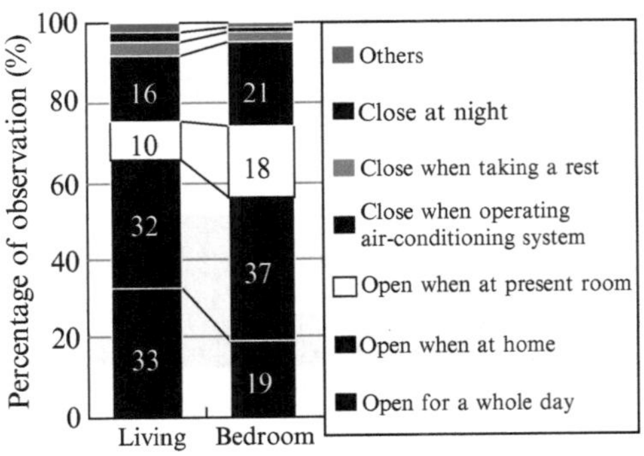

Fig. 4-47. Percentage of opening conditions

It can be noticed that utilizing natural ventilation in these areas is a common practice.

Use Patterns of Air-Conditioning Systems In Fig. 4-48, the sole use of air-conditioning systems and the use along with fan possessed a fair share of 47% in living room and fairly high at 72% in bedrooms. Furthermore, in Fig. 4-49, it is found that there was high percentage of subjects feeling uncomfortable with long periods of air-conditioning systems usage.

Awareness of Feasibility of Cross-Ventilation Due to Natural Ventilation Potential In Fig. 4-50, the percentage distribution of responses to overall comfort according to the utilization of natural ventilation is shown. From the Fig., it is found that during January to June and October to December periods, 50% of responses felt no discomfort, while approximately 20% of responses

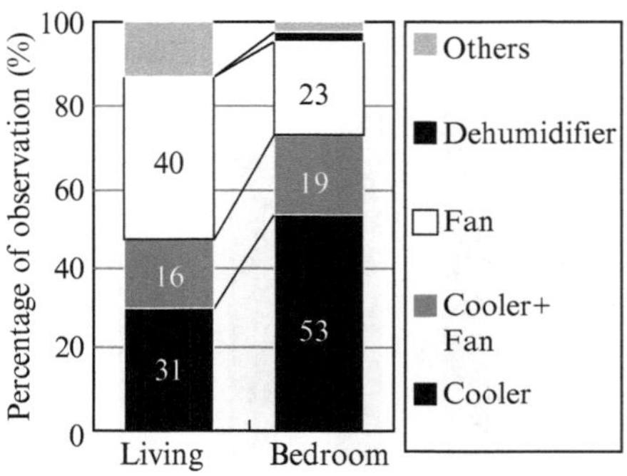

Fig. 4-48. Percentage of use pattern of air conditioning systems

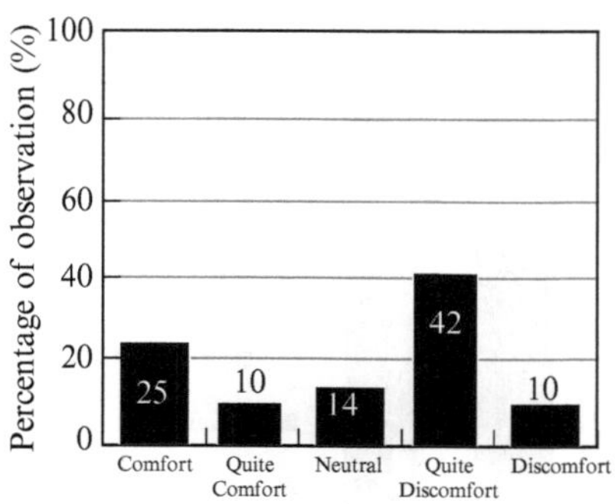

Fig. 4-49. Percentage of responses after long period of the use of air conditioning systems

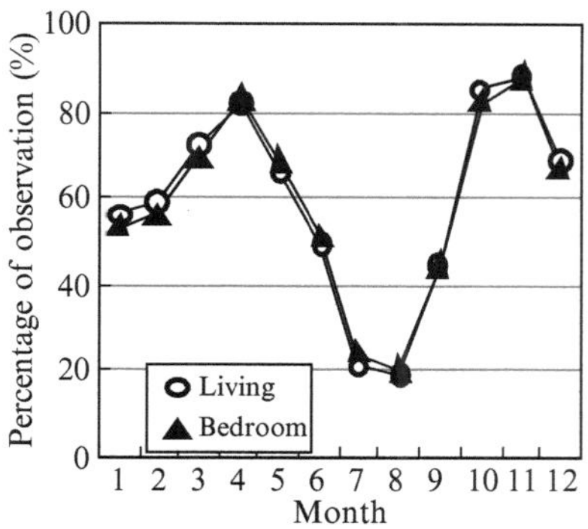

Fig. 4-50. Percentage of responses corresponding with comfort while utilizing natural ventilation

corresponded during mid-summer. Of note is the relatively high percentage of responses in semi-comfort to comfort categories. Thus, it can confirm the feasibility of cross-ventilation due to natural ventilation in Guangzhou

Indoor Airflow Driven by Natural Ventilation During Summertime As illustrated in Fig. 4-51, to the question on the preference between natural ventilation and air-conditioning systems, 80% of responses show intentions to utilize natural ventilation, which implies the preference on cross-ventilation. Additionally, in Fig. 4-52, the percentage distribution of responses to indoor air movement according to the height of subjected unit is illustrated. "Insufficient" and "Insufficient-Comfort" responses were found 63 and 52% in subjects living on first and second floors and third and fourth floors of buildings, respectively. In addition, more than half of the subjects in lower floors of buildings felt the insufficiency of air movement. These finding are similar to the results reported in the survey in Taiwan(Tu et al. 2004).

Conclusions on the Questionnaire Survey While the high expectation on cross-ventilation approach is seen, actual cross-ventilation detected in lower floors of buildings are considered to be insufficient in reality. Therefore, a solution for cross-ventilation enhancement is necessary in those cases.

4.3.3 Selection of the Type of Residences for Further Investigations

The questionnaire results collected from subjects in Guangzhou and Shenzhen in the previous section show that, the percentage of residences in either Tower or Unit type residences was above 30%. Between these two, Tower type residences

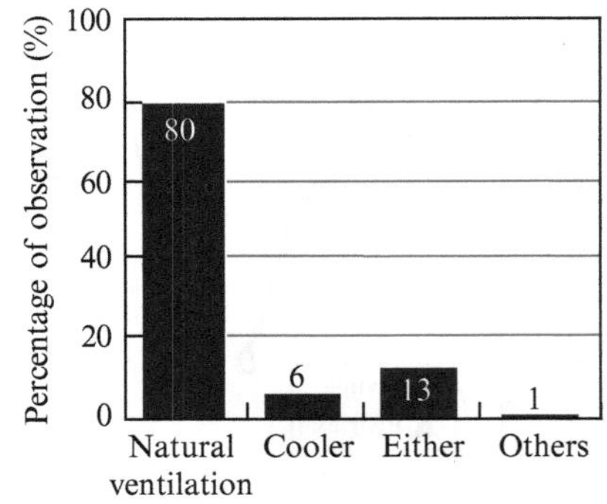

Fig. 4-51. Percentage of preferences between cooler and natural ventilation

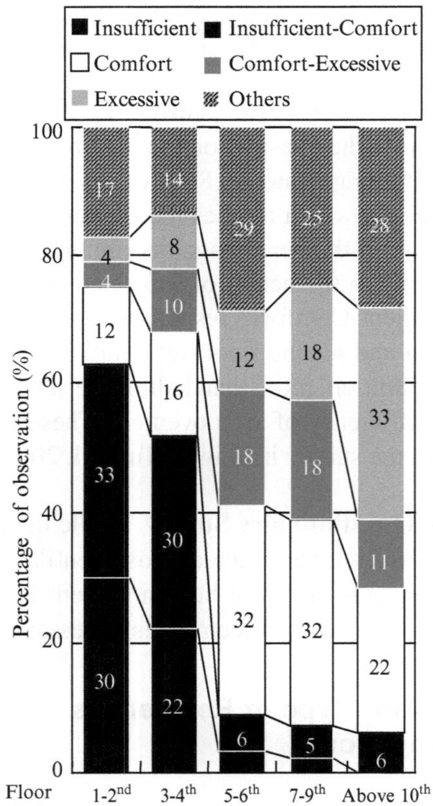

Fig. 4-52. Percentage of responses to interior airflow according to opening conditions

with relatively high percentage of high-rise buildings in particular of over ten stories are considered to possess high feasibility of cross-ventilation. However, in Unit type, which a majority of residences are mid-high-rise buildings below nine stories, the percentage of residential units below sixth floor was even higher than 70%, which provides difficulty for cross-ventilation to be realized. Therefore, the current research opts for Unit type residences for further investigations on cross-ventilation enhancement.

4.3.4 Wind Pressure Coefficient of Unit Type Buildings Measured on the Wind Tunnel Experiment

A model of a typical Unit type building was used in the wind tunnel experiment facility together with its neighborhood environment. The distribution of wind pressure coefficient was initially measured according to the ordinary configuration and later compared with that of improved configurations for cross-ventilation enhancement.

4.3.4.1 Description of Experiment

The wind tunnel experiment was conducted in the facility buildings in the University of Tokyo. A view of the building model and wind tunnel facility is illustrated in Fig. 4-53. A subjected Unit type building was modeled in 1:100 scales. The array of neighborhood buildings with two rows of buildings on the south and north sides, as well as another row of buildings on the east and west sides were also set up.

As shown in Fig. 4-54, for the purpose of reproducing the wind characteristics subjected to the group of mid-high-rise buildings, the power index representing turbulent boundary layer inlet flow was set at 0.27. Wind

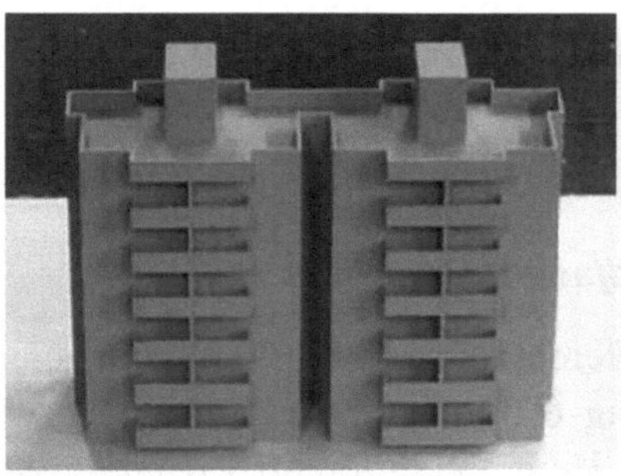

Targeted residential unit

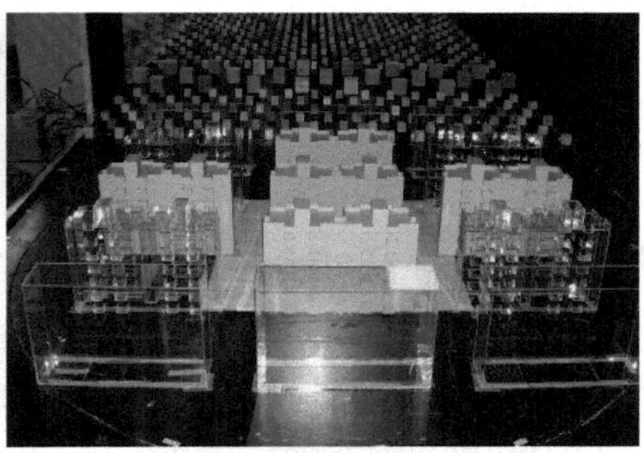

Model set up

Fig. 4-53. Model for experiment

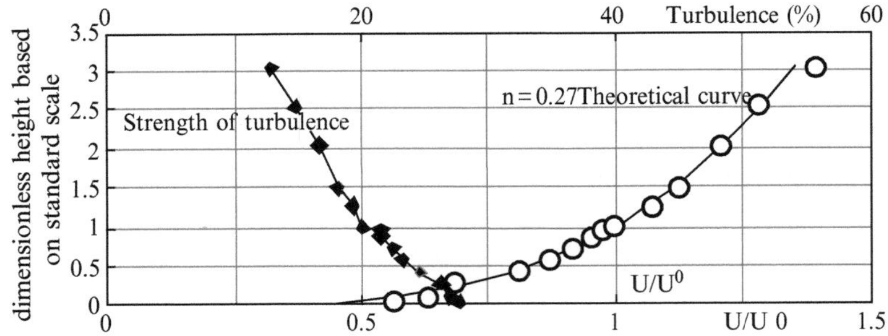

Fig. 4-54. Profile of inlet flow in wind tunnel experiment

velocity at the identical height to top level of targeted building (198 mm) was set at 7 m/s as reference wind velocity U_0. The interval of wind direction angle was set at 22.5° with 16 possible patterns.

Regarding to the typical building, as illustrated in Fig. 4-55, one building consists of four residential units with identical two units; one on the west and the others on the east. two units share a staircase acting as a core forming a left-right symmetric plan. The wind pressure coefficients were measured on two units on the east. It was assumed that openings exist at positions S1–4, N1–4 and W1–4

Experiment cases with different configurations are briefly summarized in Table 4-6. As can be seen, case 0 represented the case without any adjacent buildings, while case 1, 2, 3 differed in the interval distance with adjacent buildings on south and north at 1/4L, 1/2L and 3/4L, respectively and fixed distance at 1/2L to adjacent buildings on east and west. Furthermore, as shown in Fig. 4-56, in order to achieve the enhancement of cross-ventilation, the space between east and west units was enclosed and transformed into vertical open space from the bottom of the first floor to the building roof. Later in this paper, this element is called "void." The effect of parameters, for instance, the depth of void, total height of void, opening at first floor level etc. are further investigated, which bring the total number of experiment cases to 15.

4.3.4.2 Interval Distance Between Adjacent Buildings

In Fig. 4-57, the distribution of the differential wind pressure coefficients between position S1 and N1 comparing different interval distances with adjacent buildings in case 0–3 is shown. Here, the differential wind pressure coefficient was achieved by averaging values obtained from the tests with different 16 wind directions. By comparing results obtained from each

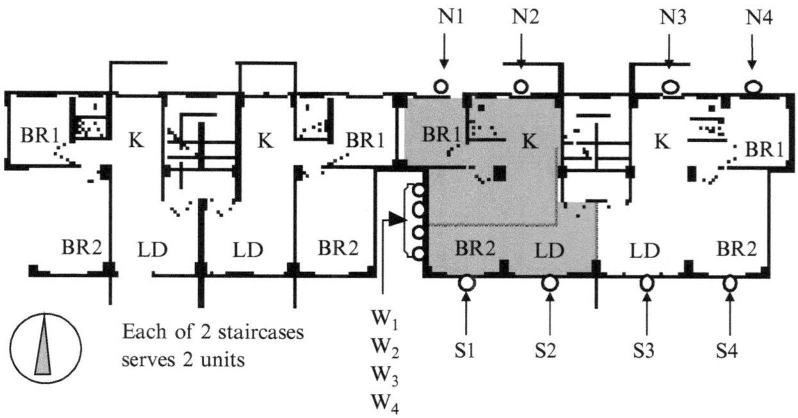

Fig. 4-55. Typical floor plan

Table 4-6. Measurement cases in wind tunnel experiment

Case	Distance to N–S adjacent buildings (*L* building height)	Void		
		Depth	Vertical extension	Opening
Case 0	None	–	–	–
Case 1	1/4L	–	–	–
Case 2	1/2L	–	–	–
Case 3	3/4L	–	–	–
Case 0–1	None	D	–	–
Case 1–1	1/4L	D	–	–
Case 2–1	1/2L	D	–	–
Case 3–1	3/4L	D	–	–
Case 1–1/4	1/4L	1/4D	–	–
Case 1–1/2	1/4L	1/2D	–	–
Case 1–3/4	1/4L	3/4D	–	–
Case 1–1–1Fa	1/4L	D	1F	–
Case 1–1–1Fb	1/4L	D	1F	Yes
Case 1-1-2Fa	1/4L	D	2F	–
Case 1–1–2Fb	1/4L	D	2F	Yes

floor, the coefficients were highest at sixth floor and gradually reduced as approaching to lower floors. As low differential wind pressure coefficient implies low cross-ventilation, it can be considered that there is a presence of the lack of cross-ventilation in low height part of buildings, which matches with the results stated in questionnaire survey. Furthermore, the effect of

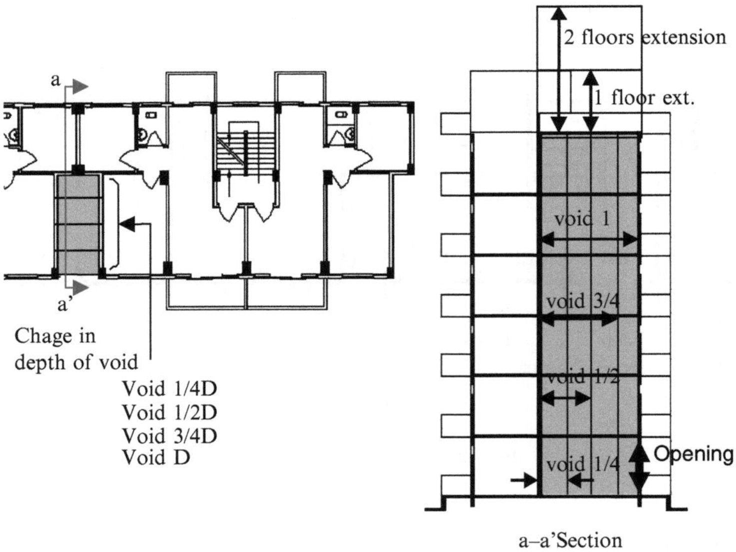

Fig. 4-56. Void configurations

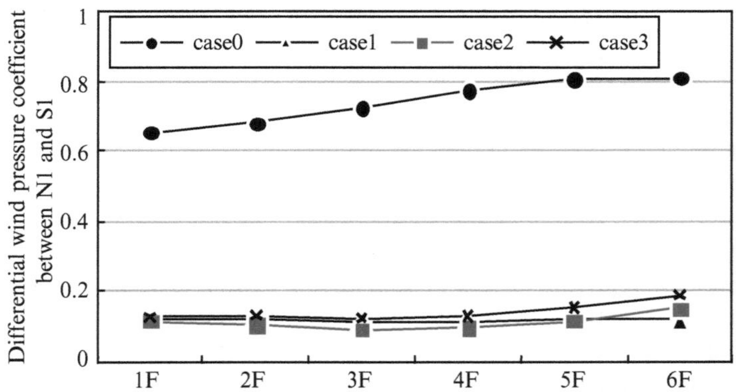

Fig. 4-57. Differential wind pressure coefficient in South–North orientation influenced by adjacent buildings

adjacent buildings on the differential wind pressure efficiency is obviously highlighted in case 1–3, which was relatively low at only 1/4 of case 0. This leads us to state that the effect of adjacent buildings is considerably significant. With previous findings in mind, a further investigation concerning void configurations to enhance cross-ventilation was performed in the following sections.

4.3.4.3 Void Basic Configurations

In Fig. 4-58, the distribution of differential wind pressure coefficient between positions exterior S1 on exterior side and W1-4 in void on fourth floor comparing the different interval distances with adjacent buildings from case 0–1 to 3–1 is illustrated. The extremely low level of the differential wind pressure coefficient can be noticed in every case including the results on other floors besides fourth floor. Thus, further investigations will only focus on the differential wind pressure coefficient between S1 and W1 instead.

Furthermore, it can be further noticed in the figure that as the interval distance with adjacent buildings reduced, the wind pressure coefficient accordingly decreased and resulted in difficulty of utilization of cross-ventilation. Thus, further investigations will pay attention to the cases with the lowest interval distance with adjacent buildings at 1/4L.

Regarding to void depth, the distributions of differential wind pressure coefficient between S1 and W1 in cases 1–1/4 and 1–3/4 are shown in Fig. 4-59. No difference in the differential wind pressure coefficient is particularly found. Thus, void depth D can be inferred as a standard depth in further investigations.

As shown in Fig. 4-60, existence of void has almost no impact on the differential wind pressure coefficient between S1 and N1. So it can be considered that void has no negative influence on the differential wind pressure coefficient in south and north directions.

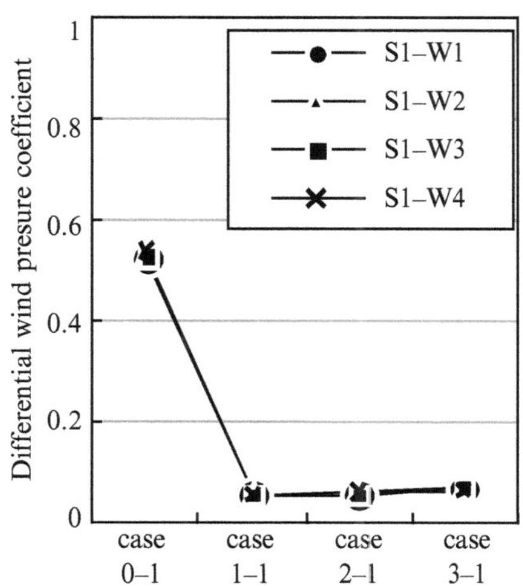

Fig. 4-58. Distribution of differential wind pressure coefficient within voids comparing different interval distances

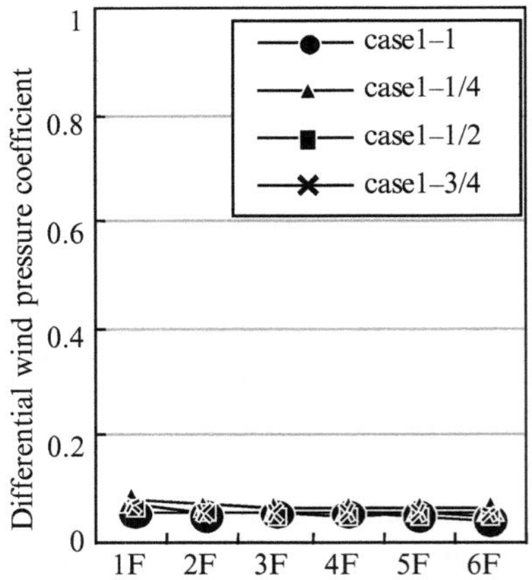

Fig. 4-59. Comparison of differential wind pressure coefficient between different void dimensions

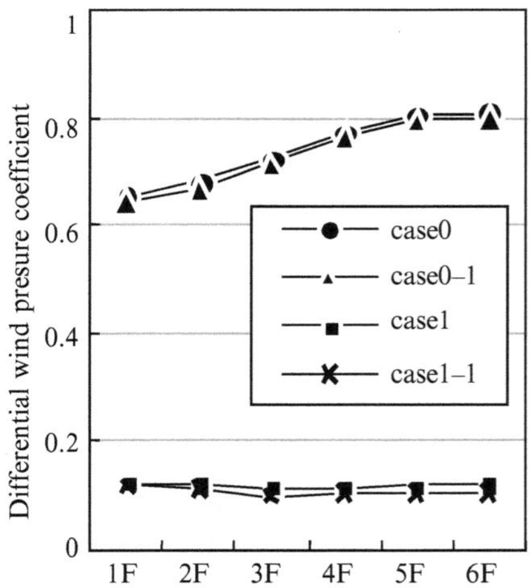

Fig. 4-60. Distribution of differential wind pressure coefficient between S1 and N1

4.3.5 Enhancement of Cross-Ventilation Due to the Presence of Voids

In Fig. 4-61, the distribution of differential wind pressure coefficient comparing cases 0, 0–1, 1, 1–1 is exhibited. From the figure, in the cases 0 and 0–1 without neighboring buildings, the presence of voids considerably

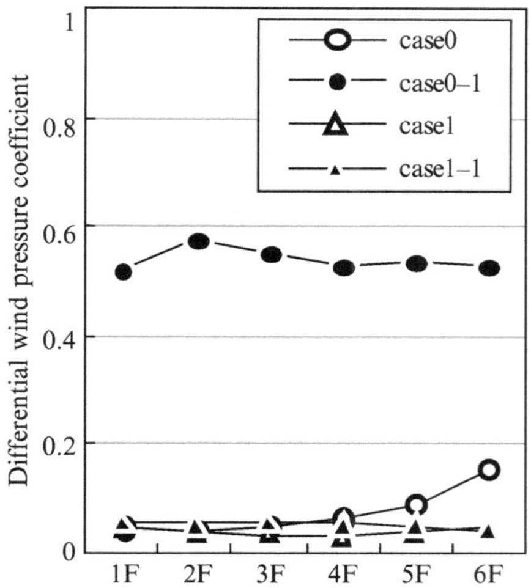

Fig. 4-61. Distribution of differential wind pressure coefficient between S1 and W1

increased the differential wind pressure coefficient and resulted in obvious enhancement of cross-ventilation. However, in the cases 1 and 1–1, which include take into account adjacent buildings into account, there is no particular influence on the differential wind pressure found. In addition to that, in an attempt to improve cross-ventilation, the extension of void height over the roof by 1 and 2 floors height in cases 1–1–1Fa and 1–1–2Fa as well as the presence of opening on first floor in cases 1–1–1Fb and 1–1–2Fb were analyzed. The results in the differential wind pressure coefficient between positions S1 and W1 of mentioned cases together with ordinary case 1 are shown in Fig. 4-62. As can be noticed, by the extension of void over the roof, the significant increase in differential wind pressure coefficient was found and consequently enhanced the cross-ventilation potential. This could be due to the fact that wind velocity generally increases as height level increases, which creates higher negative pressure over the interior side of voids. Despite of that, the placement of opening in first floor greatly reduced the differential wind pressure coefficient. In conclusion, the effects of void vertical extension are confirmed and it is suggested to take into consideration regarding to the enhancement of cross-ventilation. Furthermore, although the extension of void over the roof by 2 floors height is considered impractical, the extension over the roof by a floor height is practically possible.

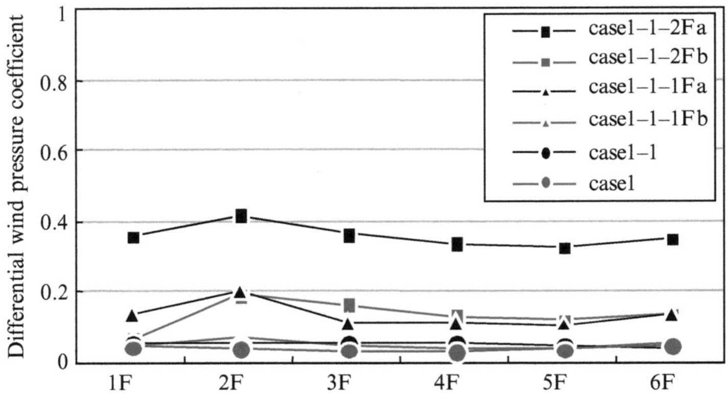

Fig. 4-62. Distribution of differential wind pressure coefficients in S1–W1 caused by different void configurations

4.3.6 Prediction of Cross-Ventilation Flow Rate in the Targeted Buildings

The cross-ventilation flow rate was estimated by using ventilation network calculation based on the wind pressure coefficients obtained from the wind tunnel experiment. The enhancement of cross-ventilation due to void configurations was therefore quantitatively evaluated.

4.3.6.1 Calculation Conditions

The targeted residential unit for calculation is illustrated as shaded area previously shown in Fig. 4-55. Additionally, flow rate calculation conditions are summarized in Table 4-7. Main doors were assumed to constantly closed. As constantly opened doors can be considered unrealistic in terms of practical occupant behaviors, a small opening over partition door was assumed for the calculation. The openings at position S1–2, N1–2 and W1 were also assumed. Flow rate coefficient was set at constant rate of 0.67 regardless of conditions involved. Reference wind velocity over the top of the building at 19.8 m above ground level was referred to the constant value of 1.84 m/s obtained by averaging values of Guangzhou's standard weather data during interim periods. In addition, 16 wind directions were involved in calculations as previously mentioned. With respect to wind pressure coefficients, the values obtained from the measurement illustrated in Table 4-7. were used as input. Additionally, regarding to the ventilation network calculation, only ventilation generated by outdoor wind was involved, while ventilation caused by temperature difference was totally neglected.

Table 4-7. Conditions for cross-ventilation calculation

Opening position	Scale of opening (m, m^2)	Windows type	Flow rate coefficient
S1	0.75 × 2.4 = 1.8	Sliding windows	0.67
S2	0.75 × 1.0 = 0.75		
N1	0.5 × 2.4 = 1.2		
N2	0.75 × 1 = 0.75		
W1	0.5 × 1 = 0.5		
Between units	0.3 × 0.9 = 0.27		

4.3.6.2 Calculation Results

In Fig. 4-63, the distribution of mean air change rate obtained by averaging the results from all 16 wind directions in the cases with the distance to adjacent buildings of 1/4L is illustrated. The air change rate of 12 ach was found in case 1, without voids. In addition, as expected, the maximum air change rate of 21 ach was obtained in the case 1–1–2Fa with voids extended over the roof by 2 floors height and without openings in first floor. This is considered to be as much as two times of air change rate obtained in case 1. In addition, in the case 1–1–1Fa with voids extended over the roof by 1 floor height obtained an air change rate of 16 ach, which is considered to be 1.3 times ordinary case 1.

4.3.7 Conclusions

In this section, the investigation on the utilization of cross-ventilation in Guangzhou and Shenzhen was mentioned. The findings are listed as follows:

1. Based on the results of SET* calculated by using the standard weather of Guangzhou, 1,850 h/year is considered to be feasible for natural ventilation utilization. Therefore, cross-ventilation can be considered as possible promising approach in this area.
2. Regarding the results of the questionnaire survey obtained from subjects in Guangzhou and Shenzhen, high awareness of natural ventilation was found. Additionally, the lack of ability to apply cross-ventilation on lower parts of buildings lead us to further investigate the solutions to enhance cross-ventilation potential.
3. Wind tunnel experiments were performed on Unit type multiple dwelling houses. The effects due to the distance with adjacent buildings and voids on wind pressure coefficient were found.

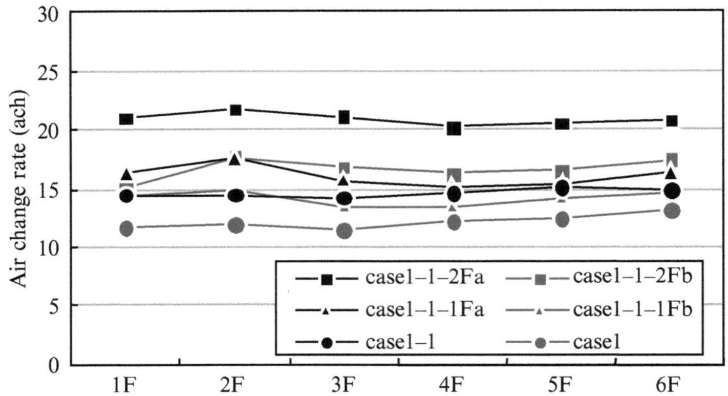

Fig. 4-63. Distribution of air change rate due to different void configurations

4. Based on the wind pressure coefficients obtained from the wind tunnel experiment, it can be considered that the presence of voids plays an important role in enhancing cross-ventilation flow rate. In addition, by using ventilation network calculation, it is possible to predict the cross-ventilation flow rate.

In summary, it can be concluded that the presence of voids without opening on the first floor of the buildings significantly improves the cross-ventilation flow rate by 1.3–2 times over the cases without voids. Furthermore, it is worth mentioning that the modifications on various aspects, for instance, staircase, placement of voids in each room, scheduled operation of opening and etc., may possibly further improve cross-ventilation due to natural ventilation.

References

Akamine Y, Kurabuchi T, Ohba M, Endo T, Kamata M (2004a) A CFD analysis of the air flow characteristics at an inflow opening. Int J Ventilation 2:431–438

Akamine Y et al (2004b) A study on evaluation of cross-ventilation performance of openings part 13: experimental study on openings with projections. Proceedings of annual meeting of the architectural institute of Japan D-2:805–806 (in Japanese)

Cai J et al (2001) Measurement of the components of Reynolds Stress and comparison of the Wind Pressure Coefficient of high-rise residential with or without a balcony with wind tunnel experiment. Proceedings of annual meeting of the architectural institute of Japan D-2:681–682 (in Japanese)

Chen R H et al (2001) A study on the assessment model of natural ventilation potential of buildings in different layout. ABRI, Taipei (in Chinese)

Chou T C (1995) The research in natural ventilation of low-rise buildings. NCKU, Tainan (in Chinese)
Chou P C (2000) A study on the naturally ventilated performances of the envelop openings. NCKU, Tainan (in Chinese)
Chou PC, Chiang CM, Chuah YK et al (1998) Effects of window positions on the air flow distribution in a cross-ventilated residential bedroom. Indoor Built Environ 7(5–6):300–307
Endo T, Kurabuchi T, Ohba M, Akamine Y, Kamata M (2004) A fundamental study on the air flow structure of outflow openings. Int J Ventilation 2:439–446
Gagge AP et al (1971) An effective temperature scale based on a simple model of human physiological regulatory response. ASHRAE Trans 77:247–262
Habara H et al. (2002) Effect of the monitor roof upon the indoor thermal environment and property of natural draft in the room: part. 7 analysis of cross-ventilation property of the monitor roof using CFD simulation, Summaries of technical papers of annual meeting of architectural institute of Japan, 755–756 (in Japanese)
Ikenoue D et al (2002) Effect of the monitor roof upon the indoor thermal environment and property of natural draft in the room: part. 3 wind tunnel test of cross-ventilation characteristics of monitor roof. Summaries of technical papers of annual meeting of architectural institute of Japan Kinki Brunch 42:293–296 (in Japanese)
Ishida Y et al (2005) Study on wind environment in urban block by CFD (part 2): wind velocity in void spaces in 3-d densely built-up area. Proceedings of annual meeting of the architectural institute of Japan D-2:825–826 (in Japanese)
Kato S et al (2005) Study on wind environment in urban block by CFD (part 1): concept of void spaces in built-up area and analysis of wind velocity over 2-d urban street block model. Proceedings of annual meeting of the architectural institute of Japan D-2:823–824 (in Japanese)
Kosmos SR, Riskowski GL, Christianson LL (1993) Force and static pressure resulting from airflow through screens. Trans ASAE 36(5):1467–1472
Li PW, Chan ST (2000) Application of a weather stress index for alerting the public to stressful weather in Hong Kong. Meteorol Appl 7:369–375
Narumi D et al (2007) Effect of the monitor roof on the indoor thermal environment and property of natural ventilation in the room. J Arch Build Sci 13(26):617–622 (in Japanese)
Shiraishi Y et al (2002) Enhancement effect of natural crossventilation and reduction effect of cooling load for porous residential building: study on reduction of environmental load by residence of urban area in hot and humid region. J Arch Plann Environ Eng 558:15–22 (in Japanese)
Tu Y et al (2003) A preliminary study in the utility of cross-ventilation in Taiwan part 1: about cumulative distribution frequency of wind speed based on climatological datum. Proceedings of annual meeting of the architectural institute of Japan D-2:607–608 (in Japanese)
Tu Y et al (2004) A preliminary study on the utility of ventilation in Taiwan dwellings part 2: the study on behavior of using ventilation and air-conditioner based

on Questionnaire Survey. Proceedings of annual meeting of the architectural institute of Japan D-2:603–604 (in Japanese)

Tu Y et al (2005) A preliminary study on the utility of ventilation in Taiwan dwellings part 3: techniques of ventilation improvement in houses of congested urban area based on the wind tunnel experiment. Proceedings of annual meeting of the architectural institute of Japan D-2:615–616 (in Japanese)

Zhang Q (2005) Development of hourly weather database and characteristics of climatic change for main Chinese cities. J Arch Plann Environ Eng 591:97–104 (in Japanese)

Zhang Q, Asano K (2001) Development of the typical weather data for the main Chinese cities. J Arch Plann Environ Eng 543:65–69 (in Japanese)

Zheng Y et al (2006) A preliminary study on the utility of cross-ventilation in southeast Asia: part1 the Question Survey in Guangzhou and Shenzhen. Proceedings of annual meeting of the architectural institute of Japan D-2:595–596 (in Japanese)

Zheng Y et al (2007) A preliminary study on the utility of cross-ventilation in southeast Asia: part2 improving cross-ventilation with a void in a TANGEN-type residential building. Proceedings of annual meeting of the architectural institute of Japan D-2:797–798 (in Japanese)

5. Advanced Monitoring of Particle-Bound Polycyclic Aromatic Hydrocarbons (pPAHs) and Risk Assessment of Their Possible Human Exposure in Roadside Air Environment in Urban Area

Tassanee Prueksasit, Kensuke Fukushi, and Kazuo Yamamoto

5.1 pPAHs Emission from Automobiles and Its Potential Risk

5.1.1 Identification of Carcinogenic PAHs in Airborne Particles and Necessity of Their Monitoring in Urban Environment

The toxic effect of most concern from exposure to PAHs is cancer. The International Agency for Research on Cancer (IARC) has classified several purified PAHs and PAH derivatives as probable (Group 2A) and possible (Group 2B) human carcinogens. In addition, the US EPA has also identified several PAHs as possibly carcinogenic to humans (Group B2). To focus a view of health effect, the cancer risk was assigned for assessing potential human exposure to pPAHs. Since the on-line monitor, photoelectric aerosol sensor (model PAS2000CE), is able to detect wide coverage of many PAHs and determine as total pPAHs concentration. It is necessary to know whether the analyzer is responsible to carcinogenic PAHs or not. Table 5-1 shows 12 PAHs that are predominantly found in particulate phase, and identified as the priority pollutants PAHs by the US EPA, and that have been classified as carcinogenic PAH by either IARC or US EPA as well. Some detectable PAHs by PAS2000CE are also included in Table 5-1.

In order to assess the potential human exposure to pPAHs from the real-time measurement, potency equivalency factors (PEFs) of some classified

H. Furumai et al. (eds.), *Advanced Monitoring and Numerical Analysis of Coastal Water and Urban Air Environment*,
© 2010 to the complete printed work by Springer, except as noted. Individual authors or their assignees retain rights to their respective contributions; reproduced by permission.

Table 5-1. List of detectable PAHs by PAS

	IARC		US EPA	Detectable	
PAH	**2A**	**2B**	**B2**	**PAH by PAS**	**PEF**
Phe				○	–
Anth				○	–
Fluor				–	–
Pyr				○	–
B(a)A	○		○	○	0.1
Chry			○	○	0.01
B(b)F		○	○	○	0.1
B(k)F		○	○	○	0.1
B(a)P	○		○	○	1.0
I(1,2,3-cd)P		○	○	○	0.1
D(a,h)A	○		○	○	0.1
B(g,h,i)P				○	

carcinogenic PAHs as given in Table 5-1, which have been developed by California EPA, were applied. Judging from the table, PAS2000CE can measure the carcinogenic PAHs, which PEF can be utilized.

5.1.2 Risk Assessment of Human Exposure to PAHs and Basis of Their Cancer Potency

Health risk assessment due to exposure to PAHs has been studied by California Environmental Protection Agency (Cal/EPA) under the "California's Toxic Air Contaminant program" (Collins and Brown 1998). For this program, B(a)P was selected as the primary representative of the class of PAHs because of (1) the relatively large amount of toxicological data available, (2) the availability of air monitoring techniques, and (3) the known and frequent human exposure to B(a)P in airborne PAHs. However, many other PAHs and PAH derivatives are also carcinogenic, the impacts of not only B(a)P but also other PAHs and PAH derivatives were considered. For risk assessment of complex pollutant mixtures, therefore, all carcinogenic PAHs have been considered to be as carcinogenic as B(a)P. Nevertheless, the possible overestimation of cancer potency was obtained by the limited number of carcinogenic PAHs included. Improving the accuracy of the risk assessment involves two parts: (1) more experimental data on the carcinogenicity of individual

PAHs and (2) more accurate estimates of the carcinogenic potency. The Office of Environmental Health Hazard Assessment (OEHHA) of the Cal/EPA, then, improved accuracy of the estimation of carcinogenicity of PAHs. Since there was no adequate information regarding the carcinogenicity of B(a)P to humans from epidemiological studies, data from animal bioassays were extrapolated to estimate human cancer risk.

Nisbet and Lagoy (1992) have proposed a toxic equivalency factor (TEF) scheme for 17 PAHs. Nevertheless, several of the 17 PAHs had not been declared carcinogens by either IARC or US EPA. The US EPA (1999) issued a relative potency for PAHs, which are classified as carcinogens (Group 2B). Likewise, the OEHHA has developed a potency equivalency factor (PEF) for assessing the impact of carcinogenic PAHs in ambient air. Some equivalent factors are given in Table 5-1. The OEHHA did not develop PEFs for noncarcinogenic PAHs because they are not currently identified as carcinogens by authoritative bodies. Cancer risk associated with exposure to ambient levels of B(a)P was estimated by extrapolating from the experimental data to ambient levels. The US EPA used the data for respiratory tract tumors from inhalation exposure in hamsters to estimate cancer potency and unit risks associated with exposure to BaP (OEHHA 1999). Because of the limited amount of data currently available for risk assessment of BaP, the inhalation unit risk of 1.1×10^{-3} $(\mu g/m^3)^{-1}$ is used as the best value for inhalation exposure (OEHHA 1999).

5.1.3 Emission Factors of pPAHs from Diesel Engine Vehicles

Emission factor of pPAHs is a function of various factors including engine type, air/fuel mixture ratio, emission control, load, age, fuel type and driving mode, including cold starting. PAH emission amount can then vary widely. In Bangkok, particulate matter is the major air pollution concern and is largely contributed by diesel vehicles. Thus, a laboratory experiment of diesel vehicle PAH emission factors was conducted (Nilrit et al. 2005). Test vehicles in this study are categorized into two groups, heavy-duty diesel vehicles (HDV) and light-duty diesel vehicles (LDV). All testing vehicle characteristics are shown in Table 5-2. The vehicle types and obtained emission factors of 16 PAHs (US EPA priority pollutants) are presented in Tables 5-3 and 5-4. In the study, gas phase and size-fractioned particulate phase PAH emissions were investigated. As shown in the table, heavy-duty diesel engines, especially buses had high emission factors.

Table 5-2. Characteristic of test vehicles in 2004

Vehicle type	Mileage	Brand	Manufactured year (age)	Sample code
Heavy duty diesel				
Bus				
(A) Mileage > 300,000 km	944,697	ISUZU	1989 (15)	BusA1
	650,565	HINO	1988 (16)	BusA2
	809,304	HINO	1997 (7)	BusA3
(B) Mileage < 300,000 km	294,058	HINO	1991 (13)	BusB1
	292,239	ISUZU	2002 (2)	BusB2
	197,532	HINO	1988 (16)	BusB3
	68,930	ISUZU	1989 (15)	BusB4
(C) Unidentified mileage	–	HINO	1988 (16)	BusC1
	–	HINO	1967 (37)	BusC2
	–	HINO	1988 (16)	BusC3
Bus: EURO-II	66,964	HINO	2004 (1)	EURO-II
Bus: NGV	–	BENZ	–	NGV
Truck	108,867	VOLVO	2002 (2)	Truck
Light duty diesel				
Pick-up	317,658	NISSAN	1994 (10)	PUA1
(A) Mileage > 300,000 km	396,002	ISUZU	1995 (9)	PUA2
	562,194	TOYOTA	1990 (14)	PUA3
	626,549	TOYOTA	1990 (14)	PUA4
(B) Mileage < 300,000 km	44,400	TOYOTA	2003 (1)	PUB1
	85,405	ISUZU	2001 (3)	PUB2
	176,070	ISUZU	2000 (4)	PUB3
	197,527	TOYOTA	1996 (8)	PUB4
	209,452	TOYOTA	1996 (8)	PUB5
Van	292,805	TOYOTA	1993 (11)	VAN1
	310,715	TOYOTA	1993 (11)	VAN2
	295,171	TOYOTA	1993 (11)	VAN3

5.2 Temporal and Spatial Variation of pPAHs in Roadside Environment in Tokyo, Japan and Bangkok, Thailand

5.2.1 Diurnal Profile of pPAHs Concentration

The temporal variation of the pPAHs concentration was observed by real-time monitoring using PAS2000CE. In general, the daily change profiles of pPAHs studied in either Tokyo or Bangkok were similar. A sharp increase in pPAHs concentration was found in early morning with the peak concentration observed during 7:00 to 8:00 a.m., and followed by a significant

Table 5-3. Emission factors of total 16 PAHs of heavy duty diesel vehicles

Sample phase	**Emission factor of total PAHs (mg/km) Heavy duty diesel vehicles (HDDV)**					
	BUSA	BUSB	BUSC	EURO-II	NGV	Truck
Particle in different size range						
>18	69.28	38.34	13.27	0.97	1.74	1.30
10–18	55.91	54.16	16.14	1.34	0.36	0.25
5.6–10	61.37	57.97	14.43	1.44	0.56	0.36
3.2–5.6	64.22	63.24	24.01	0.81	0.40	0.68
1.8–3.2	54.64	42.34	27.34	1.42	1.29	0.60
1.0–1.8	71.87	49.86	29.08	2.74	1.07	0.69
0.56–1.0	79.96	73.90	29.54	1.55	4.27	0.67
0.32–0.56	64.71	104.24	38.10	2.45	1.43	1.28
0.18–0.32	62.15	106.58	39.65	1.95	1.06	2.01
<0.18	50.89	136.29	125.36	4.57	3.07	3.75
Total particulate phase	635.00	726.92	356.92	19.25	15.25	11.59
Gas phase	2.14	2.86	3.44	3.79	3.46	3.34
PM and gas phases	637.14	729.78	360.36	23.03	18.71	14.93

Table 5-4. Emission factors of total 16 PAHs of light duty diesel vehicles

	Light duty diesel vehicles (LDDV)		
Sample phase	PUA	PUB	VAN
Particle in different size range			
>18	2.04	0.25	1.52
10–18	1.97	0.23	0.26
5.6–10	1.06	0.43	0.28
3.2–5.6	0.98	0.38	0.31
1.8–3.2	3.07	0.79	2.10
1.0–1.8	1.06	0.67	1.86
0.56–1.0	1.15	1.76	5.01
0.32–0.56	2.83	2.43	4.99
0.18–0.32	2.97	3.73	4.92
<0.18	8.61	14.36	24.89
Total particulate phase	25.74	25.02	46.15
Gas phase	0.43	0.17	0.17
PM and gas phases	26.18	25.19	46.33

and rapid reduction during the daytime and gradually increased once again in the evening. More explicit and higher peak concentrations in the morning than that of the general areas were presented at roadside areas. This indicates that the originated morning peak pPAHs concentrations strongly correspond to the traffic growth in the morning rush hours. This event is similar to those seen in large cities such as Basel, Switzerland, and Aosta, Italy (Chetwittayachan et al. 2002a, b, c).

The pPAHs concentration profiles at each sampling site in Tokyo in summer and in winter are displayed in Fig. 5-1a, b and Fig. 5-2, respectively. In winter, the daily change profile at the ground level (TRS3, 1.5 m height) at roadside area was not clear as observed in summer. This is probably due to the location change closer to the source than the previous observations at TRS1 (16.5 m height and 10 m far from the main street) and TRS2 (8.5 m height and 15 m far from the main street). Considering from available data at the same sampling site, i.e. TGA1 (18 m height and 100 m far from the

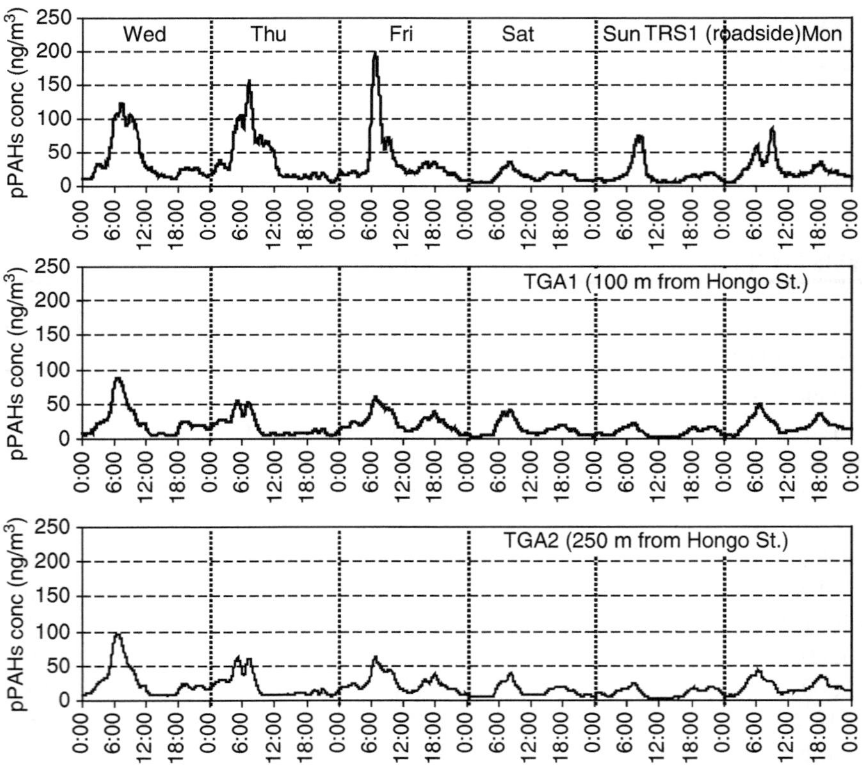

Fig. 5-1a. Diurnal profiles of pPAHs concentration in summer, during: (**a**) August 23–28, 2000, (**b**) September 19–25, 2001

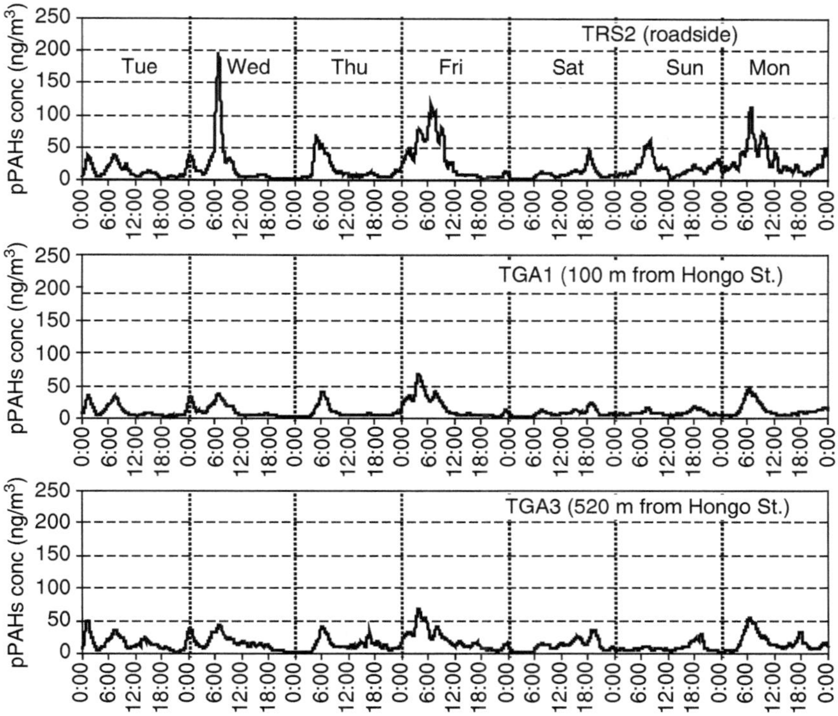

Fig. 5-1b. Diurnal profiles of pPAHs concentration in summer, during: (**b**) September 19–25, 2001

main street), the profiles gave almost similar pattern in particular on fine days. Table 5-5 shows the pPAHs concentration of weekdays in both measurements was significantly higher than that on weekends possibly due to a higher traffic amount on those days.

In Bangkok, the pPAHs concentration was measured at all sampling sites in hot and wet seasons as illustrated in Fig.5-3a, b. Throughout the week observed of both periods, the profiles on weekends were similar when compared to those of weekdays. The profile at the general area (BGA site, 15.2 m height and 200 m far from the main street) during the latter measurement shows relatively higher pPAHs concentrations than that of the former result. In addition, during the second observation, the profile of indoor pPAHs concentration at the BID site (13.2 m height) in relation to those at outdoor, in particular at the BRS1 (1.5 m height and 16.8 m far from the main street) and BRS2 (13.2 m height and 22.6 m far from the main street), was also shown in Fig. 5-3b.

Various average pPAHs concentrations of the 24-h, 6-h of morning period, and 8-h of nighttime are reported in Table 5-6. At the BRS1 site,

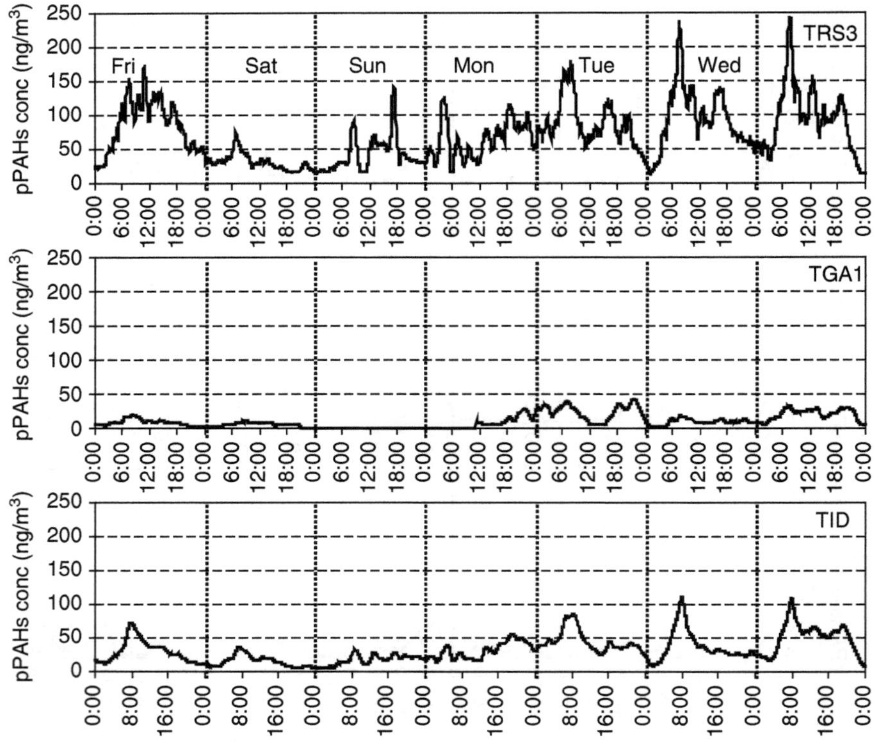

Fig. 5-2. Diurnal profiles of pPAHs concentration in winter, during January 26–February 1, 2001

Table 5-5. Average pPAHs concentration (ng/m^3) over the periods of measurements in summer in Tokyo (WD: weekday, WE: weekend)

Sampling	24-h		6-h (4:00–10:00)		8-h (16:00–24:00)	
Site	WD	WE	WD	WE	WD	WE
August	36	16	78	29	19	14
TRS1	22	11	41	20	17	12
TGA1	23	12	43	19	18	13
TGA2						
September	23	15	52	20	10	18
TRS2	13	8	26	9	7	11
TGA1	17	11	29	9	11	15
TGA3						

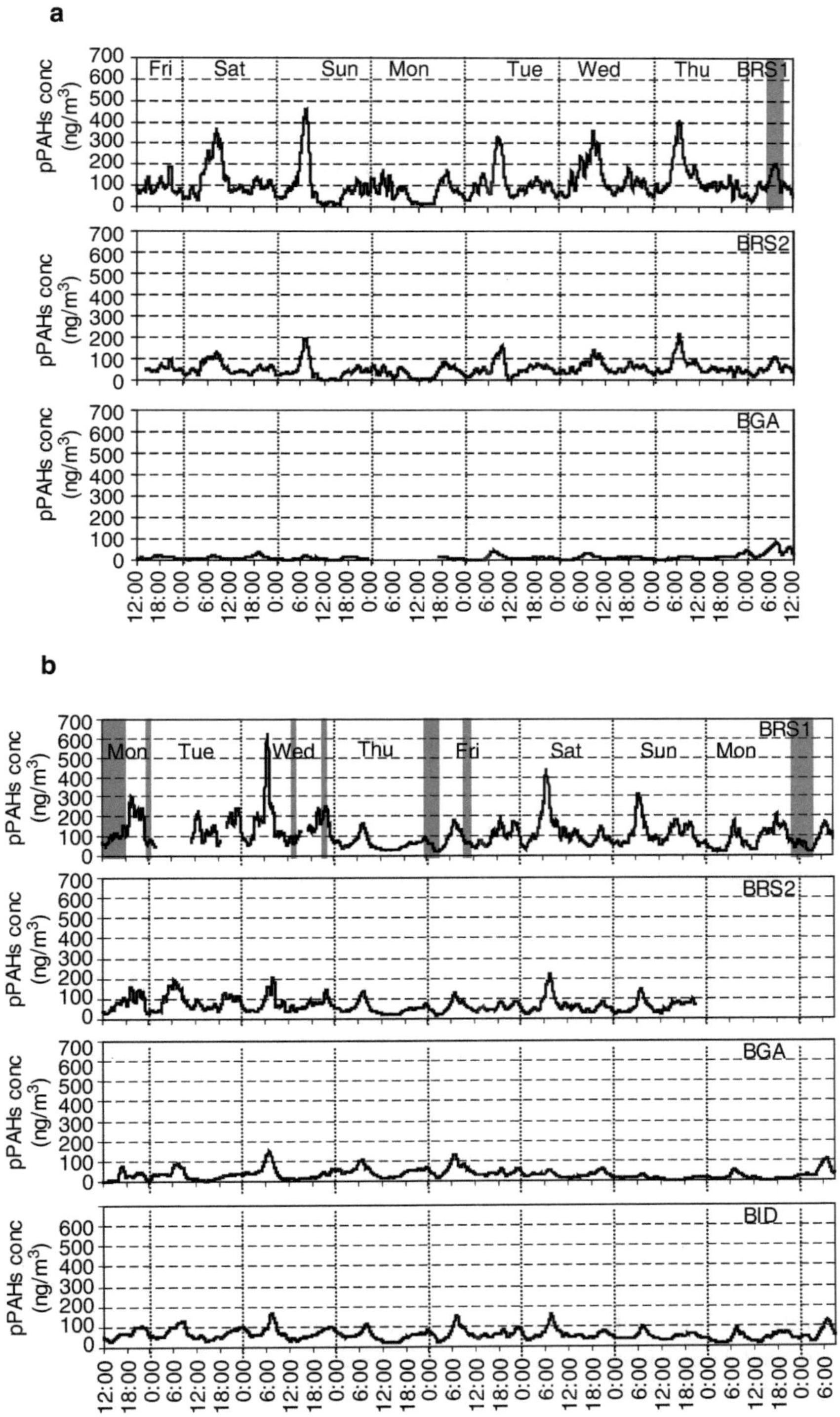

Fig. 5-3. Diurnal profiles of pPAHs concentration in hot and wet season: (**a**) March 2–9, 2001, (**b**) August 6–14, 2001 (The period of rainfall is shadowed)

Table 5-6. Average pPAHs concentration (ng/m^3) over the periods of measurements in hot and wet season in Bangkok. (WD: weekday, WE: weekend)

Sampling	24-h		6-h (4:00–10:00)		8-h (16:00–24:00)	
site	WD	WE	WD	WE	WD	WE
March						
BRS1	107	111	180	229	97	87
BRS2	54	50	78	95	52	45
BGA	17	11	22	12	14	17
August						
BRS1	108	117	128	191	134	99
BRS2	71	63	100	94	77	64
BGA	36	20	63	26	32	22
BID	62	54	82	77	67	51

although the morning average and the nighttime average in both periods were varied, the 24-h average concentrations over the observations period were similar. On the other hand, the pPAHs concentrations at the BRS2 and BGA sites during the wet season were relatively higher than those in the hot season, giving the ratios of the average concentration of the former to that of the latter of 1.3:1, and 2:1, respectively.

5.2.2 Spatial Variation of pPAHs Concentration

In view of spatial variation observed in Tokyo in summer, for the first monitoring period, the average pPAHs concentrations at both the TGA1 and TGA2 sites, where located with the distances of 100 and 250 m from the Hongo Street, respectively, were almost the same as summarized in Table 5-5. The difference between the averages of pPAHs concentrations at the both sites was statistically insignificant (at 5% significance level). This indicates that the dispersion of pPAHs at such a scale exhibited relatively uniform. On the other hand, the pPAHs concentration at the roadside was about 1.5 times higher than that at both the TGA1 and TGA2 sites. In the case of the second monitoring, the difference of pPAHs at the TGA1 and TGA3 (10 m height and 520 m far from the main street) sites was statistically significant, giving the ratio of average concentration at the TGA1 site to the TGA3 site 0.8. Hence, the location at TGA3 was near to other main roadsides, the pPAHs concentration must not only be affected by the road-originated pPAHs from Hongo Street, but also other pPAHs sources, e.g. incineration facilities, surrounding this monitoring site. The spatial variation pattern of the second sampling period showed a bit difference. The pPAHs concentration at roadside was 1.8 and 1.4 times higher than that at TGA1 and TGA3, respectively.

In case of Bangkok, considering the sampling sites at the same level, the pPAHs concentration at roadside (BRS2) in the hot and wet seasons was respectively 3.3 and 2.2 times higher than that at general area (BGA). This reflects that pPAHs pollutant more diffused during in wet season. The correlation between the concentrations at ground floor at a height of 1.5 m (BRS1 site) and at a height of 13.2 m from the ground level (BRS2 site) during both sampling periods was determined. A good agreement between the concentrations at both levels could be obtained as the R^2 values of 0.88 in hot season and of 0.72 in wet season, as illustrated in Fig. 5-4a, b, respectively. From these figures, the decrease of the average pPAHs concentration from the BRS1 site to the BRS2 site can be easily calculated by the equations of the relationship between the concentration at ground floor and four-storey height level.

5.2.3 Comparison of pPAHs Concentrations in Tokyo and Bangkok

In order to compare the results between both cities, it is desirable to consider similarity of sampling site location and meteorological condition. The data sets in summer during the sampling period of August, 2000 (Tokyo),

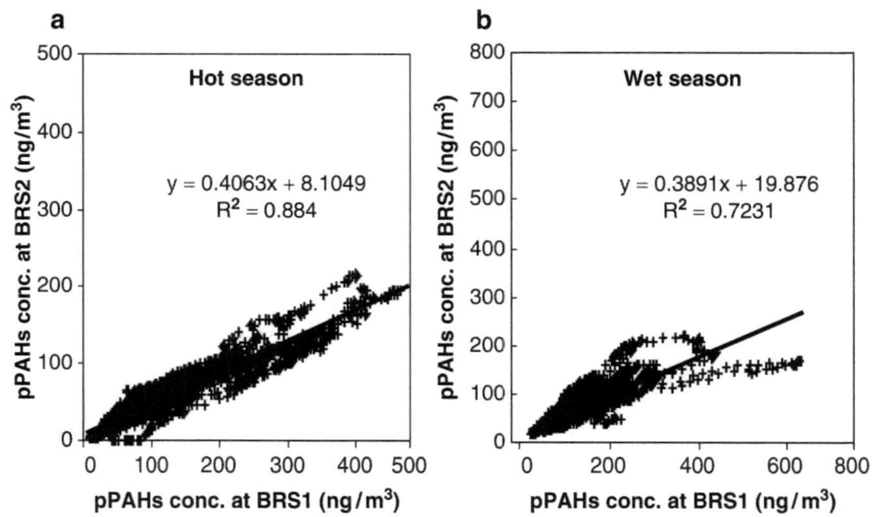

Fig. 5-4. (**a**, **b**) Relationship between pPAHs concentrations at both the BRS1 and BRS2 sites in hot and wet season

and of March, 2001 (Bangkok), as given in Tables 5-5 and 5-6, were then selected. The TRS1 and TGA2 sites in Tokyo, and the BRS2 and BGA sites in Bangkok were considered to represent roadside and general areas.

Average pPAHs concentrations shown in the table were focused for 24-h average, morning peak, and nighttime concentrations. At roadside areas, all the average concentrations over the period of observation in Bangkok were significantly higher than that of Tokyo; especially, the nighttime concentration in Bangkok was about 2.8 times higher than that of Tokyo due to larger amount of traffic flow. Meanwhile, the concentrations observed at the general area of Tokyo were slightly higher than those in Bangkok. No large difference in 24-h average pPAHs concentrations were observed between roadside and general area in Tokyo, whereas, larger difference could be found in Bangkok, giving the ratio of the average concentration at roadside to that of the general area a value of 3.3:1. This suggests that widespread dispersion of pPAHs occurred over the entire study area of the city of Tokyo, and indicates that the potential risk associated with human exposure to pPAHs within general areas is higher in the city of Tokyo than that in Bangkok.

5.2.4 Relationship Between Indoor and Outdoor pPAHs Concentrations

The result studied in Tokyo reveal that significant quantitative differences between outdoor (TRS3) and indoor (TID) pPAHs concentrations were observed as reported in Table 5-7. The daily profiles of pPAHs concentration in the indoor area, which located near the road without major indoor source of the pollutant such as smoking, gave similar trend to those at roadside. A satisfactory correlation between outdoor and indoor pPAHs concentrations could be obtained giving the Pearson correlation value of 0.85 or the R^2 value of 0.71, as illustrated in Fig. 5-5. This indicates that pPAHs concentration in indoor air corresponded well to that in outdoor air.

From Table 5-7, both mean and median of either indoor or outdoor pPAHs concentrations were assessed. As a result, indoor-outdoor ratios of pPAHs concentration determined by either mean or median during the

Table 5-7. Average pPAHs concentrations measured at indoor and outdoor

Location	Mean	SD	Median
Indoor	32.26	20.40	27.33
Outdoor	70.84	42.32	64.15
Indoor: outdoor ratio	0.46		0.43

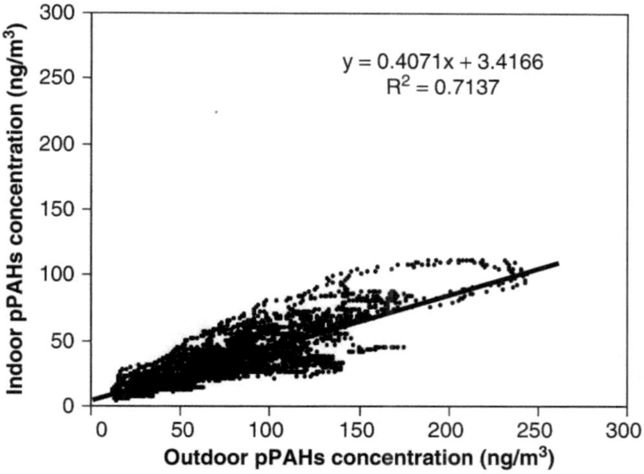

Fig. 5-5. Relationship between indoor and outdoor pPAHs concentration in Tokyo

whole sampling period were similar, giving the ratio of average concentration at indoor to outdoor of 0.46 and 0.43, respectively.

In Bangkok during wet season, the measurement of indoor pPAHs concentration (BID site) at the same level as the BRS2 site was conducted. A correlation between indoor and outdoor concentrations either at the BRS1 and BRS2 site was examined and the results are shown in Fig.5-6a, b. At the same level of measurement, a linear correlation could be obtained with the R^2 value of 0.49. A linear correlation between indoor and outdoor pPAHs concentration at the ground floor was observed with the R^2 value of 0.54. In addition, at the same receptor height, indoor-outdoor ratios of various average pPAHs concentration, i.e. for 24-h, morning peak period, and nighttime, gave similar values of 0.87, 0.82, and 0.86, respectively. This indicates that throughout a day the ratio of indoor to outdoor concentrations was relatively constant. Additionally, people spending the time at this indoor site would expose to almost the same pPAHs level as in the outdoor air environment. When comparing to the results in Tokyo, a relatively higher of the indoor-outdoor ratio of 0.87 observed in Bangkok, while in Tokyo, the value of 0.46 was observed. A higher indoor-outdoor ratio found in Bangkok because the air ventilation in the indoor site of Bangkok was better than that of Tokyo. In case of no direct indoor source, risk assessment of human exposure to pPAHs at the indoor area can be easily estimated by utilizing the correlation from Fig. 5-6.

Some studies associated with indoor and outdoor pPAHs concentrations have been revealed. For instance, Kingham et al. (2000) determined the concentration of traffic-related total PAHs absorbed on PM_{10} in indoor and

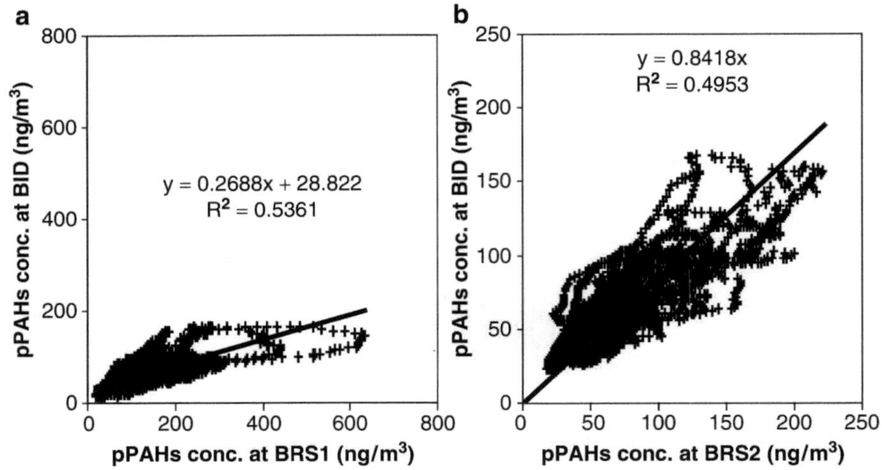

Fig. 5-6. (**a**, **b**) Relationship of pPAHs concentrations at the BID site to those either at the BRS1 and BRS2 site

outdoor air in Huddersfield, UK. They found that, for 'proximity' homes (roadside), indoor concentrations were about 0.65 of outdoor concentrations, and the indoor-outdoor concentration ratio of background homes gave the value of 0.50. Moreover, for homes defined on the basis of their modelled NO_2 concentration (one 'high NO_2' and one 'low NO_2'), indoor-outdoor ratio of total PAHs concentration were assessed as 0.50 and 0.57 for high and low NO_2 homes, respectively. Data documented by Fischer et al. (2000) give indoor:outdoor ratio of 0.57 and 0.59 for total PAHs concentrations in PM_{10} at high and low traffic density homes, respectively, in Amsterdam. Ando et al. (1996) concluded that the concentration of B(k)F, B(a)P, and B(g,h,i)P adsorbed on fine particles of <2 μm aerodynamic diameter in indoor air increased in proportion to those in outdoor air around a main road and in residential areas of Tokyo and Beijing. Even different method used and different particle size interested, the result from the real-time measurements in this study focused similar. The result is consistent with previous studies with significantly lower indoor than outdoor pPAHs concentrations for inside the room with no obvious indoor source.

In indoor area with no obvious indoor sources of pPAHs, diffusion of pPAHs from outdoor sources must be contributing to indoor pPAHs pollution. Therefore, indoor pPAHs concentration would correspond to that at outdoor with a time delay required for transportation of pPAHs from outdoor to indoor. To determine time lag between outdoor and indoor pPAHs concentration time series, the cross-correlation was estimated. In general, the cross-correlation is the correlation of a series with another series, shifted by a particular number of observations (StatSoft 1999). The computation of cross-correlation function (C_{xy}) can be expressed as:

$$C_{xy}(\mathrm{k}) = E[(x_t - x_m)(y_{t+x} - y_m)] \tag{5.1}$$

where C_{xy} is the cross-correlation function between the time series of x and y, k is the number specified in the number of lags, x_t is the x value at time t, y_{t+k} is the y value at time t with shifting forward for a lag of k, x_m *and* y_m are the mean values of x and y time series, respectively. The cross-correlation coefficient (R_{xy}) is computed by the following equations:

$$R_{xy}(k) = \frac{E[(x_t - x_m)(y_{t+x} - y_m)]}{\sigma_x \sigma_y} \tag{5.2}$$

where

$$\sigma_x = \mathrm{E}[(x_t - x_m)^2] = variance(x) \tag{5.3}$$

$$\sigma_y = \mathrm{E}[(y_{t+x} - y_m)^2] = variance(y) \tag{5.4}$$

Lag	R_{xy}	Std.err
−60	.5302	.0142
−55	.5525	.0142
−50	.5752	.0142
−45	.5985	.0141
−40	.6224	.0141
−35	.6469	.0141
−30	.6722	.0141
−25	.6993	.0141
−20	.7289	.0141
−15	.7595	.0141
−10	.7899	.0141
−5	.8188	.0141
0	.8448	.0141
5	.8607	.0141
10	.8673	.0141
15	.8653	.0141
20	.8554	.0141
25	.8383	.0141
30	.8155	.0141
35	.7904	.0141
40	.7640	.0141
45	.7370	.0141
50	.7098	.0142
55	.6831	.0142
60	.6568	.0142

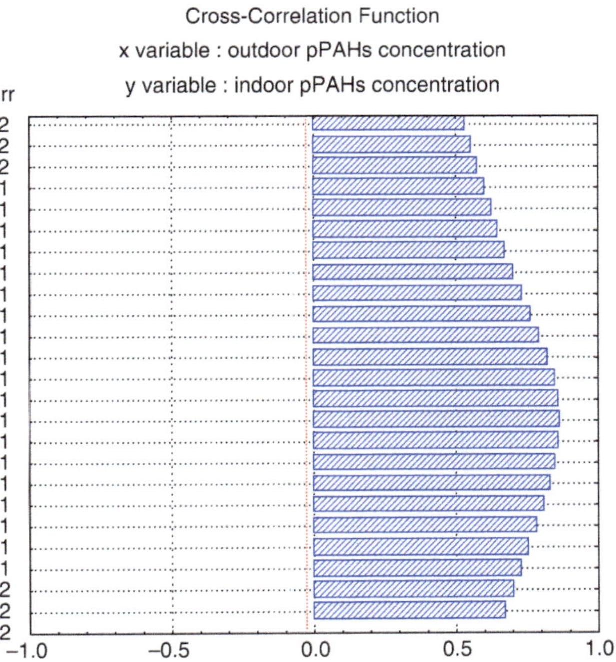

Fig. 5-7. Cross-correlation coefficient between outdoor and indoor pPAHs concentration measured in Tokyo

Time series data of both outdoor and indoor pPAHs concentrations at 2-min interval measured by PAS2000CE in Tokyo and in Bangkok were taken to determine the cross-correlation. Figure 5-7 shows the cross-correlation coefficient of the data obtained from Tokyo at a lag of -*k* to +*k*. The highest cross-correlation coefficient (R_{xy} = 0.8673) at the lag of 10 could be obtained. Thus, the transportation of pPAHs from outdoor to indoor was quantitatively shown with the corresponding time delay of 20 min.

In case of Bangkok, the cross-correlations between outdoor pPAHs concentrations at 2-min intervals at either the ground level (BRS1 site) or a height of 13.2 m from the ground (BRS2 site) and indoor pPAHs concentration at BID site were determined. There were strong cross-correlations (R_{xy} = 0.7057 and R_{xy} = 0.8308, Fig. 5-8a, b) between indoor pPAHs concentrations and those at the BRS1 and BRS2 sites with the lags of 12 and of 8, respectively. As a result, the transportation of pPAHs from outdoor at the ground level and at the same height as indoor to indoor area were shown with the corresponding time delay of 24 and 16 min, respectively.

5.2.5 Periodic Component of pPAHs

A study on dynamics of pPAHs has not been done due to lack of the pPAHs time series data. Because of the suitability of the PAS2000CE for pPAHs time series screening, it is desirable to investigate their dynamics and some considerable influences. Therefore, the measurement of pPAHs concentration over the period of 2 months was undertaken at 2-min interval at the TRS1 site, see Fig. 5-9.

To explore cyclical pattern of pPAHs time series data, spectrum analysis was performed. Theoretically, the purpose of the analysis is to decompose a complex time series with cyclical components into a few underlying sinusoidal (sine and cosine) functions of particular wavelengths (StatSoft 1999). The "wave length" of a sine and cosine function is typically expressed in terms of the number of cycles per unit time (Frequency:ν). In addition, the period (T) of a sine and cosine function is defined as the length of time required for one full cycle. Thus, it is the reciprocal of the frequency, or: $T = 1/\nu$ (StatSoft 1999). The general structural model can be expressed as:

For k = 1 to q

$$X_t = a_0 + \sum [a_k * \cos(\lambda * t) + b_k * \operatorname{sine}(\lambda * t)] \tag{5.6}$$

where λ is the frequency expressed in terms of radians per unit time, that is: $\lambda = 2\pi * \nu$. Note that the cosine parameters a_k and sine parameters b_k are

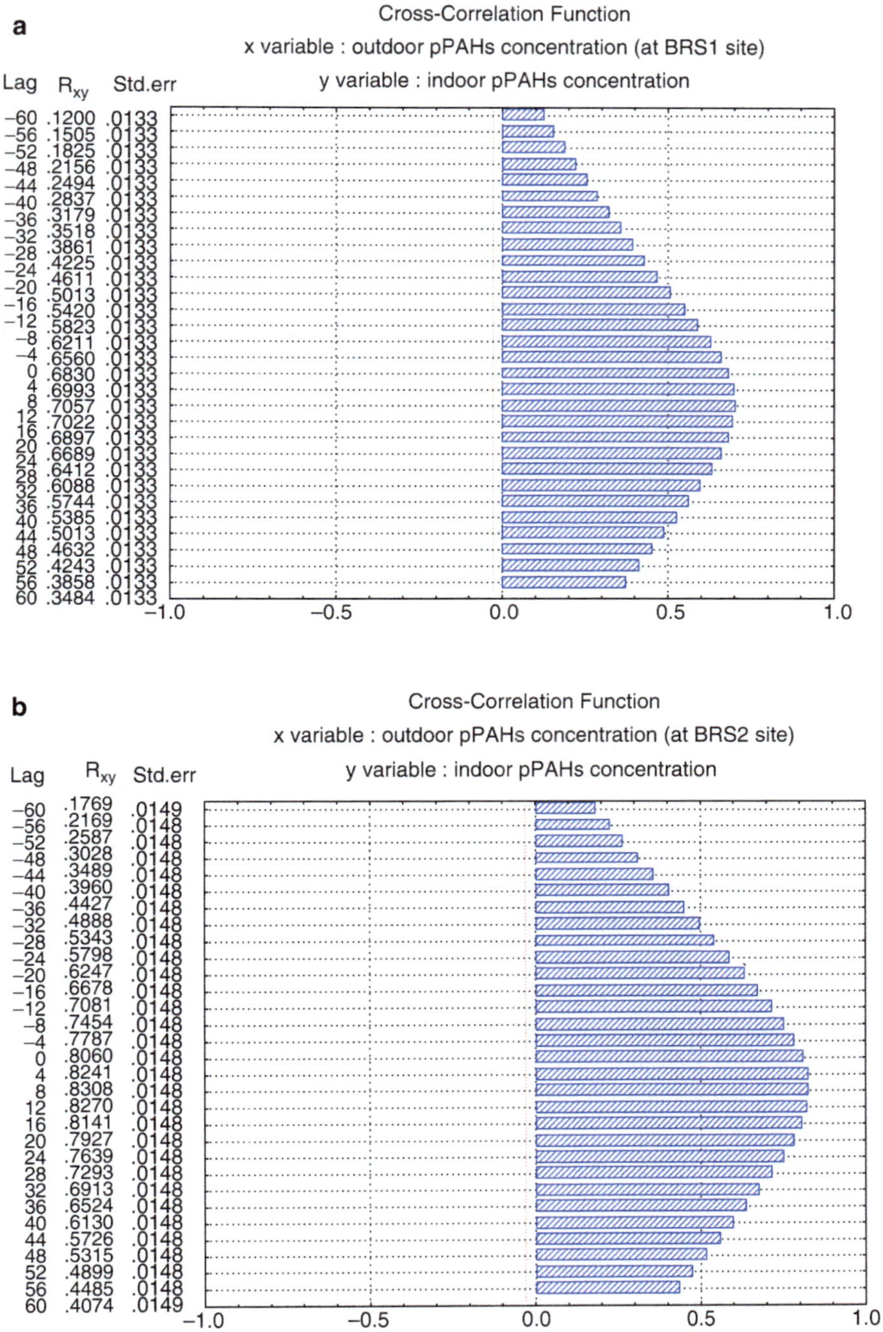

Fig. 5-8. Cross-correlation coefficient between outdoor pPAHs concentration at the BRS1 (**a**) and BRS2 sites (**b**) and those in the indoor site

coefficients that tell the degree to which the respective functions are correlated with the data. Overall there are q different sine and cosine functions.

The sine and cosine functions are mutually independent; thus sum of the squared coefficients for each frequency can provide the periodogram as the following computation:

$$P_k = (b_k^2 + a_k^2) * n / 2 \tag{5.7}$$

where P_k is the periodogram value at frequency v_k and n is the overall length of the series. Generally, the periodogram values themselves are subject to substantial random fluctuation. A clearer picture of underlying periodicities often only emerges when examining the spectral densities. In order to do so, the spectral density estimates are computed by smoothing the periodogram values with a weighted moving average. The Blackman-Tukey smoothed periodogram, i.e. Hamming window, was used because it is a consistent spectral estimator (Hies *et al.* 2000).

Figure 5-9 shows the original time series of pPAHs concentration during the whole observation period. The first data point corresponds to 18 May 2001 (Friday), and the last one to 15 July 2001 (Sunday). The data are clearly fluctuating. Interestingly, the data show a stationary long-term component, where it was likely to be no systematic change in mean.

pPAHs time series, which consist of 42,480 data points, were analyzed in the frequency space with spectra of periodogram. Figure 5-10a, b present the periodogram values and the spectral density of pPAHs concentration, which was plotted against frequency, respectively. The largest amplitudes of a spectrum indicate the main periodicities of the underlying processes. From Fig. 5-10, the large peaks present in the specific frequency range of 0–0.005, and this range covers the approximately 320 frequency points of the total 7,999 points. To obtain significant peaks, testing for white noise series in certain frequency ranges was carried out, and the numbers of the frequency points were divided into four ranges: 0–40, 41–80, 81–160, and 160–320.

After considering criterion of significance of those peaks, some significant peaks in Fig. 5-10 can be obtained at the following frequencies: 0.0014, 0.0028, 0.0002, 0.0012, 0.0006, and 0.0042 cycle per 2-min. These frequencies correspond to the periods of 1, 0.5, 7, 1.2, 2.3, and 0.3 [d], respectively.

The results reflect that at the TRS1 site, where located at roadside area, three predominant peaks, i.e.1 and 0.5 [d] periodic component of pPAHs concentration and a weekly pattern, could be identified for all series. Three additional peaks of 1.2, 2.3, and 0.3 [d] could be also detected visually in this series.

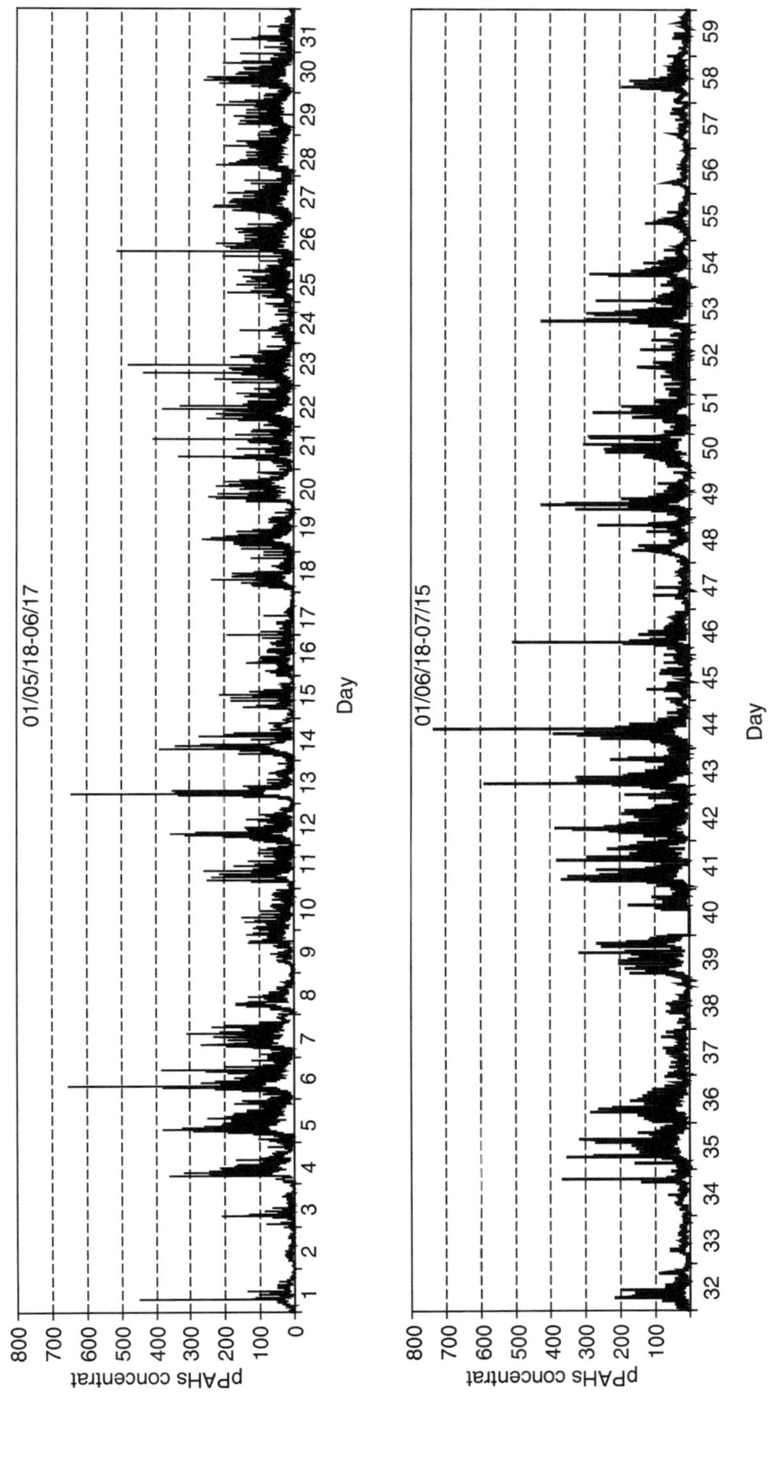

Fig- 5-9. pPAHs concentration time series at the TRS1 site in Tokyo during the period of May 18 to July 15, 2001

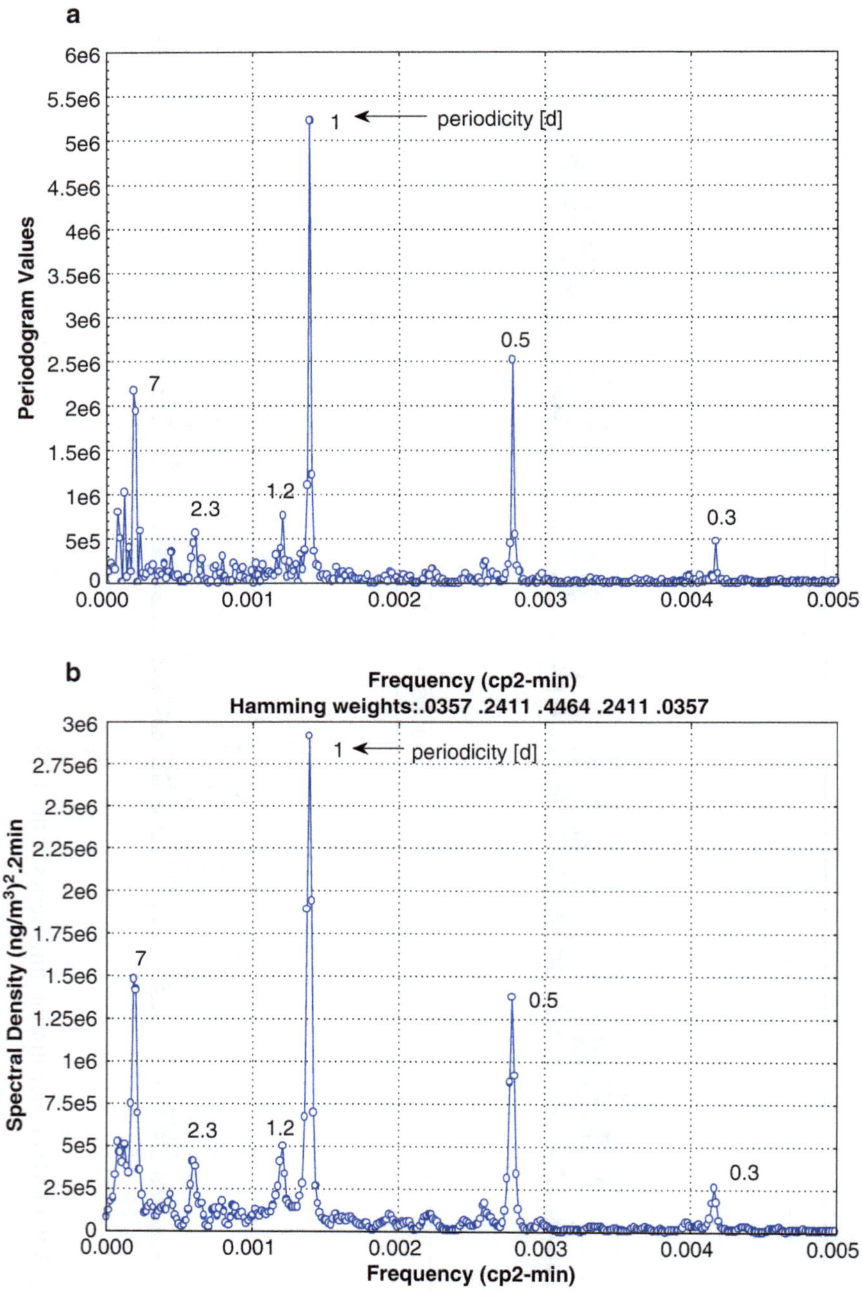

Fig. 5-10. Periodogram value (**a**) and spectral density (**b**) for the pPAHs 2-min interval at the TRS1 site. The frequency is in cycles per 2 min

5.3 Risk Estimation of Potential Human Exposure to pPAHs

5.3.1 Integrating On-Line and Off-Line Measurement for Risk Estimation of Human Exposure to pPAHs

Particle-bound PAHs concentration is typically determined by traditional chemical analysis methods, including gas chromatography/mass spectrometry (GC/MS), high-performance liquid chromatography (HPLC), which are likely to be costly and time consuming for continuous air pollution screening purpose. To obtain continuous temporal variation of pPAHs concentrations, in particular for fine particle, the use of a device capable of giving real-time detection of pPAHs concentration like Photoelectric Aerosol Sensor (PAS) appears attractive. Nevertheless, this technique provides only the sum of PAHs concentration, without giving information on individual PAH species. Otherwise, the relationship between gas chromatographic chemical analysis of filter extracts with the PAS signal output should be verified. There are some remarkable results those revealed that linear relationship between the total concentration of PAHs adsorbed on the particles and PAS output could be obtained over a large range of concentrations (Agnesod et al. 1996; McDow et al. 1990; Siegmann and Siegmann 1998). It should be noted that overestimation of human exposure risk might be considerable if we directly assess it based on the total PAHs concentration detected by the PAS alone. However, once PAS detection can be converted to carcinogenic PAHs determined by the traditional methods, it would offer rather simple methodology to estimate potential human exposure to pPAHs. Therefore, on-line continuous monitoring response with off-line measurement using GC/MS was investigated in order to utilize for risk estimation of potential human exposure to pPAHs in Tokyo and Bangkok.

The total concentration of the selected 12 PAHs determined by GC/MS which followed the US EPA method (US EPA 1999) (as given in Table 5-1) and the average of total pPAHs concentration output from PAS2000CE during the both samplings in Tokyo and in Bangkok were plotted to examine the correlation. Satisfactory linear correlations could be obtained with the R^2 value of 0.543 for Tokyo and of 0.723 for Bangkok as shown in the Fig. 5-11.

The total of PAHs adsorbed on the particulate surface measured with PAS technique was greater, by about one order of magnitude, than the sum of the twelve PAHs selectively determined with off-line technique. Agnesod et al. (1996) also reported that the same difference in the order of magnitude was observed between off-line sampling technique using either high or low

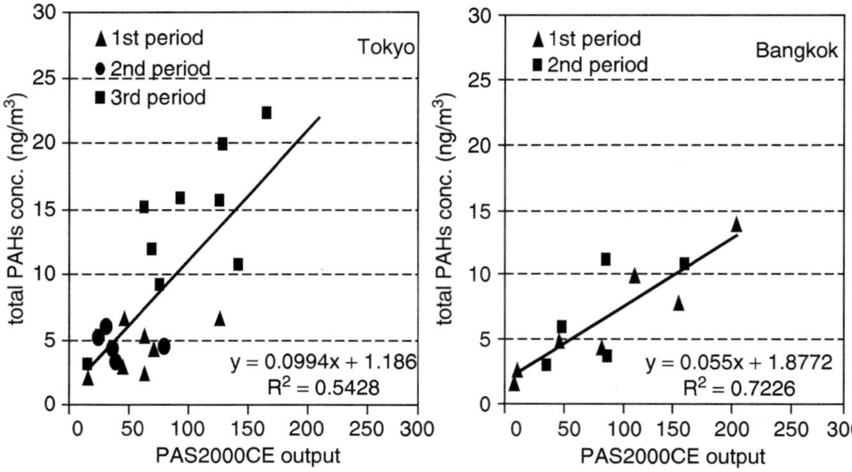

Fig. 5-11. Relationship between PAS2000CE output and total concentration of the PAHs selectively determined in Tokyo and Bangkok

volume air sampler and the automatic analyzer based on a photoelectric aerosol sensor (PAS), having determined 10 PAHs, with four or more rings, in airborne particulate which was collected in urban area.

When considering the 12 PAHs predominantly found in particulate phase (see Table 5-1), the PAS2000CE can detect 11 PAHs, including Phe, Anth, Pyr, B(a)A, Chry, B(b&k)F, B(a)P, I(1,2,3-cd)P, D(a,h)A, and B(g,h,i)P. Therefore, the composition of 11 PAHs absorbed on particles collected either in Tokyo or in Bangkok was selected to compare PAS2000CE output and GC/MS results. Figure 5-12 shows the average of percentage abundance of pPAH content in the particulate samples by the off-line technique. The contribution of detectable PAHs by PAS2000CE in the air particles in both cities studied was almost the same. Comparing the total percentage of seven detectable carcinogenic PAHs (B(a)A, Chry, B(b&k)F, B(a)P, I(1,2,3-cd)P, and D(a,h)A) in Tokyo and in Bangkok, the contributions were similar, i.e. 68% and 66% of the total PAHs respectively.

The total concentration of the 11 PAHs was calculated and was plotted against the average of total pPAHs concentration measured by PAS2000CE over each same sampling period, as presented in Fig. 5-13. A satisfactory correlation between the total of selected 11 PAHs concentration and PAS2000CE output could be acquired with the values of R^2 of 0.541 and 0.709 for Tokyo and Bangkok, respectively.

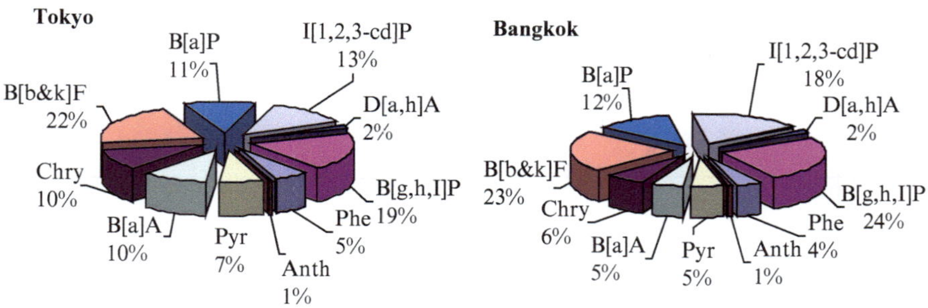

Fig. 5-12. Percentage abundance of the 11 detectable PAHs by PAS2000CE in airborne particulate

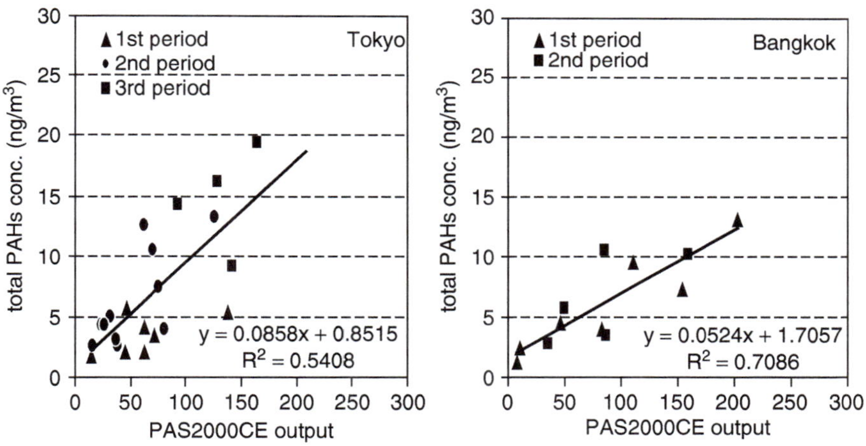

Fig. 5-13. Relationship between PAS2000CE output and total concentration of the eleven PAHs

5.3.2 Application of PEF Scheme

The concentrations of the seven carcinogenic PAHs were weighted by their PEFs (as reported in Table 5-1), and were plotted with the PAS2000CE output. The correlation between total PEF-weighted concentration (ng/m^3) and total pPAHs concentration measured by the on-line monitor in Tokyo and in Bangkok could be obtained with the R^2 values of 0.588 and 0.627, respectively, as illustrated in Fig. 5-14.

As a result, there are two possible ways to apply the relationship between carcinogenic PAHs and total pPAHs concentration; (1) directly using the correlation between total PEF-weighted concentration and total pPAHs

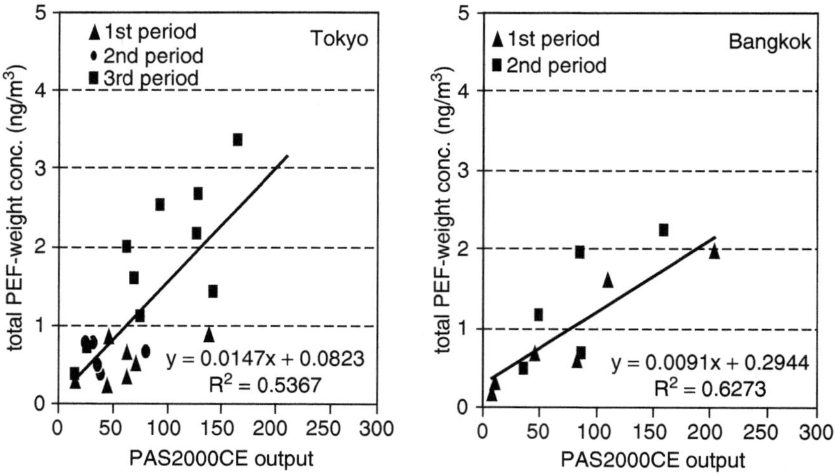

Fig. 5-14. Relationship between total PEF-weighted concentration of the carcinogenic PAHs and PAS2000CE output

concentration, and (2) applying the relationship between total concentration of the 11 PAHs detectable by PAS2000CE and total pPAHs concentration followed by utilizing the percentage abundance of the carcinogenic PAHs. For the former scheme, carcinogenic PAHs concentration from the Tokyo and the Bangkok data can be calculated by the following expressions;

$$\textit{Total PEF-weight conc.} = 0.0147\ (\textit{PAS output}) + 0.0823 \qquad \text{(5.8) (Tokyo)}$$

$$\textit{Total PEF-weight conc.} = 0.0091\ (\textit{PAS output}) + 0.2944 \qquad \text{(5.9) (Bangkok)}$$

For the latter application, from Fig. 5-13, the regression correlation of both Tokyo and Bangkok results can be utilized as below;

$$\textit{Total PAHs conc.} = 0.0858\ (\textit{PAS output}) + 0.8515 \qquad \text{(5.10) (Tokyo)}$$

$$\textit{Total PAHs conc.} = 0.0524\ (\textit{PAS output}) + 1.7057 \qquad \text{(5.11) (Bangkok)}$$

Next, the total PAHs concentration is converted to the total PEF-weight concentration by the following formula;

$$\textit{Total PEF-weight conc.}\ (ng/m^3) = \sum_{i=1}^{7} (PEF_i \ x\ F_i)\ x\ TPAHs \qquad (5.12)$$

where PEF_i is the PEF of carcinogenic PAH_i, F_i is the percentage of carcinogenic PAH_i of the eleven PAHs, and TPAHs is total concentration of the eleven PAHs.

5.3.3 Cancer Risk Estimation of Potential Human Exposure to pPAHs

The Office of Environmental Health Hazard Assessment (OEHHA) of the California Environmental Protection Agency (Cal/EPA) has developed a potency equivalency factor (PEF) for assessing the impact of carcinogenic PAHs in ambient air. Due to the limited amount of data currently available for risk assessment of B(a)P, the inhalation unit risk of 1.1×10^{-3} $(\mu g/m^3)^{-1}$ is used as the best value for inhalation exposure. The concentrations of carcinogenic PAHs are weighted by their PEFs, then, the lifetime cancer risk can be estimated as the following expression (Chetwittayachan et al. 2002a, b, c):

$$\textit{Cancer risk} = \textit{PEF-weighted conc.}\ (ng/m^3)\ x\ \textit{conversion factor}\ (0.001 \mu g/ng)\ x\ \textit{exposure time ratio}\ x\ \textit{unit risk}\ (\mu g/m^3)^{-1} \tag{5.13}$$

5.3.3.1 Risk Estimation of Potential Human Exposure to pPAHs in Tokyo

From both observations in summer (August and September) and in winter (January), the hourly average pPAHs concentration of weekdays and weekend (Fig. 5-15), and of the whole sampling period (Fig. 5-16), were calculated. All the profiles are depicted in Figs. 5-15 and 5-16 respectively. The lifetime cancer risk is, therefore, estimated by;

$$\text{Cancer risk} = \left(0.0147 \frac{\int_{t_1}^{t_2} C dt}{t_2 - t_1} + 0.0823 \right) \times 0.001 \times \frac{\text{T}_{\text{exp}}}{24} \times 1.1 \times 10^{-3} \tag{5.14}$$

where C is the pPAHs concentration determined by PAS2000CE at time t, T_{exp} is the exposure time period.

Considering that TGA1 site represented as the general area, and that the pPAHs concentrations at this site were observed in all measurements, the cancer risk of the people at this site was then considered as background risk. Consequently, the cancer risk at the TRS1, TRS2, TRS3, and TID sites were estimated as additional lifetime cancer risk due to road traffic. The results of the additional cancer risk of potential human exposure to pPAHs in the vicinity of road in Tokyo on weekday and weekend are summarized in Table 5-8, and for the whole period as shown in Table 5-9.

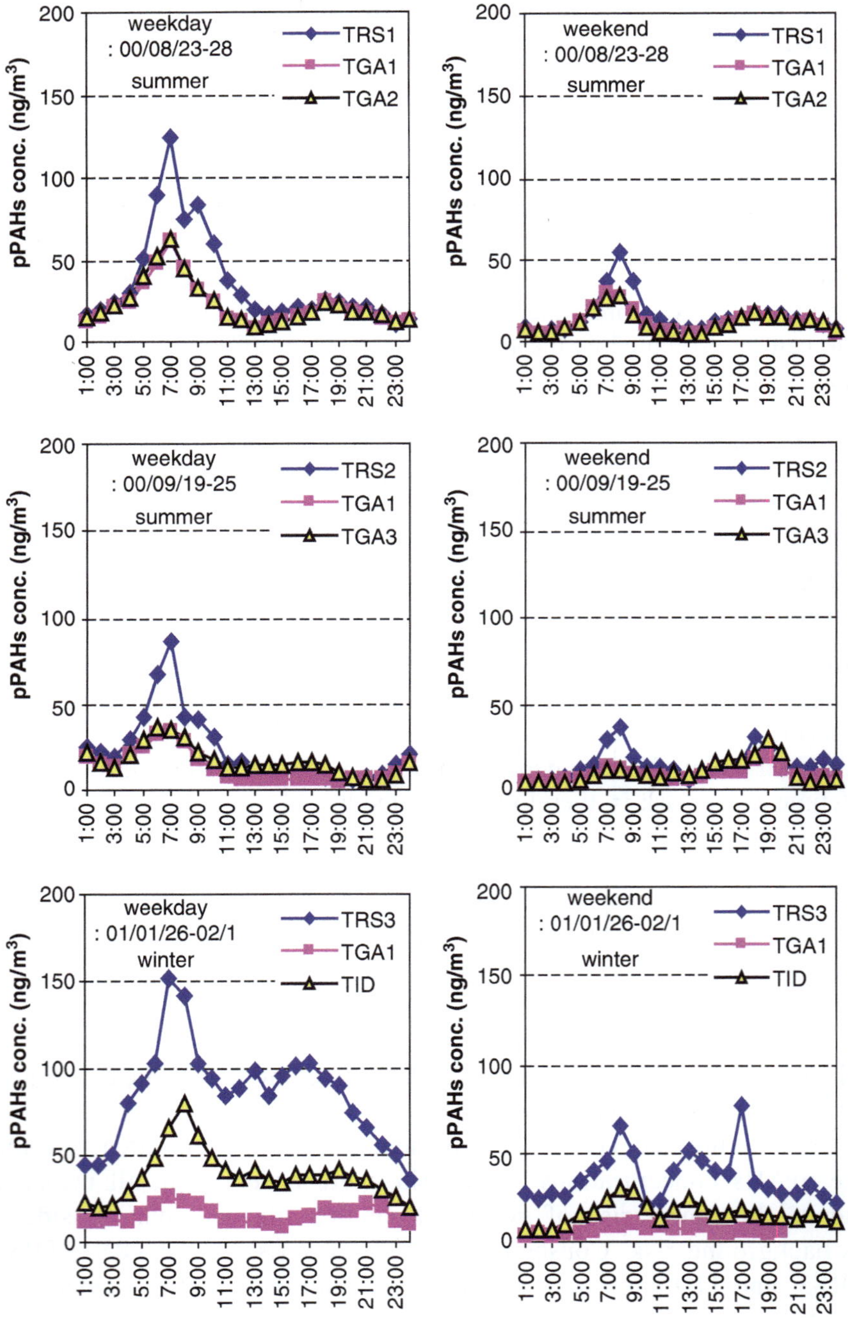

Fig. 5-15. Profile of hourly average pPAHs concentration of weekday and weekend during summer and winter in Tokyo

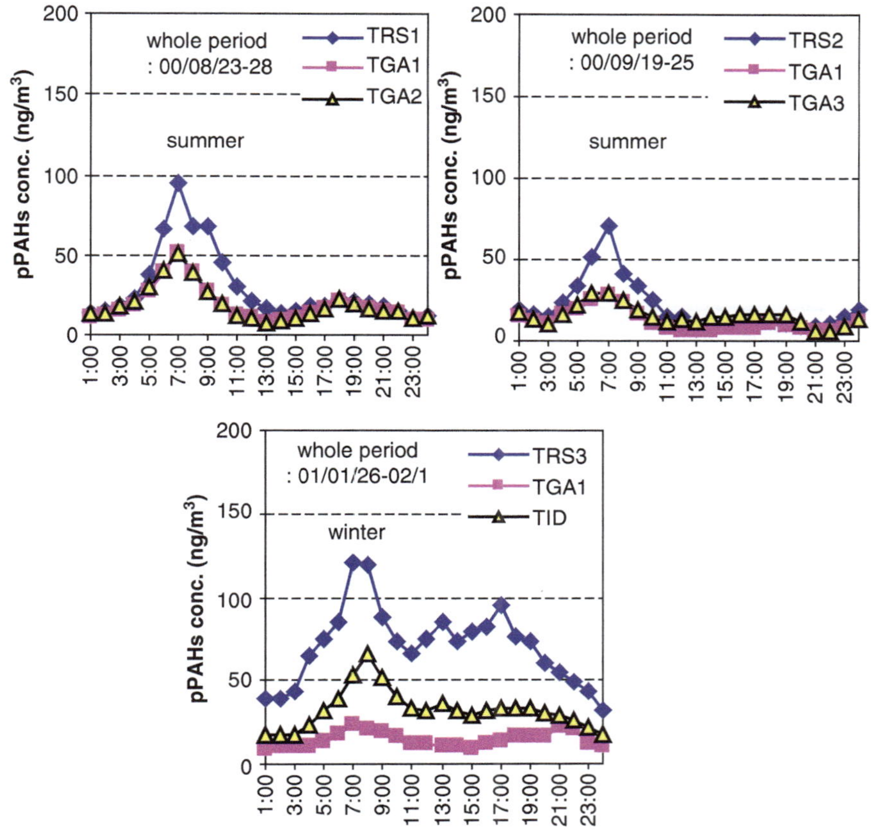

Fig. 5-16. Profile of hourly average pPAHs concentration of the whole sampling periods during summer and winter in Tokyo

Table 5-8. Additional lifetime cancer risk of potential human exposure in the vicinity of road in Tokyo on weekday and weekend

	August		September		January	
	WD	WE	WD	WE	WD	WE
Background risk						
General area (TGA1)	4.4E-07	2.7E-07	3.0E-07	2.3E-07	3.4E-07	1.7 E-07
Additional risk						
Roadside	3.2E-07 (TRS1)	1.6E-07 (TRS1)	2.5E-07 (TRS2)	1.9E-07 (TRS2)	1.2E-06 (TRS3)	5.4 E-07 (TRS3)
Indoor (TID)					4.7E-07	2.4 E-07

Table 5-9. Additional lifetime cancer risk of potential human exposure in the vicinity of road in Tokyo during whole sampling period

	August	September	January
Background risk			
General area (TGA1)	3.8E-07	2.8E-07	3.2E-07
Additional risk			
Roadside	2.7E-07 (TRS1)	2.4E-07 (TRS2)	1.0E-06 (TRS3)
Indoor (TID)			3.9E-07

From Table 5-8, the results show that the additional risk of the people at the roadside area during workdays was about two times higher than that on weekend. Among all the results, the highest additional cancer risk could be found at the TRS3 site during the observation in winter, and it was 2–2.5 times higher than those in summer. This is probably due to the location of the TRS3 site that is closer to the source than the other sites.

The background cancer risks estimated during observation periods at the general area, the TGA1 site, were almost similar in the range of $2.8–3.8 \times 10^{-7}$. From this result, the exposure of the people to pPAHs at this general area did not change even in different seasons. Table 5-9 shows that an additional risk at the roadside due to the road traffic was estimated as 2.5×10^{-7} in summer. In winter, the cancer risk of the street people at the ground level, if any, was four times higher than that at general area. When comparing between indoor and outdoor at roadside area, the additional risk estimated at outdoor area was 2.5 times higher than that at indoor area.

From Fig. 5-16, the hourly average profile of pPAHs concentration clearly showed the peak during the morning period. It is desirable to know the percentage contribution of the additional cancer risk during that period, especially at roadside area. The profiles of pPAHs at the roadside during the whole sampling periods either in summer and in winter were then subdivided into three periods, including morning (4:00–12:00), daytime (12:00–20:00), and nighttime (20:00–4:00). From Fig. 5-17, during a day, the additional risk during the morning period accounts for 40–50% of the total. While the additional risk during the daytime and nighttime distributed for 26–38% and 22–28% of the total, respectively. It is noted that a pPAHs emission control (e.g. motor vehicles inspection, maintenance program, and gasoline/diesel engine vehicles control program) against road traffic in the morning must be effective for a significant reduction of the additional risk.

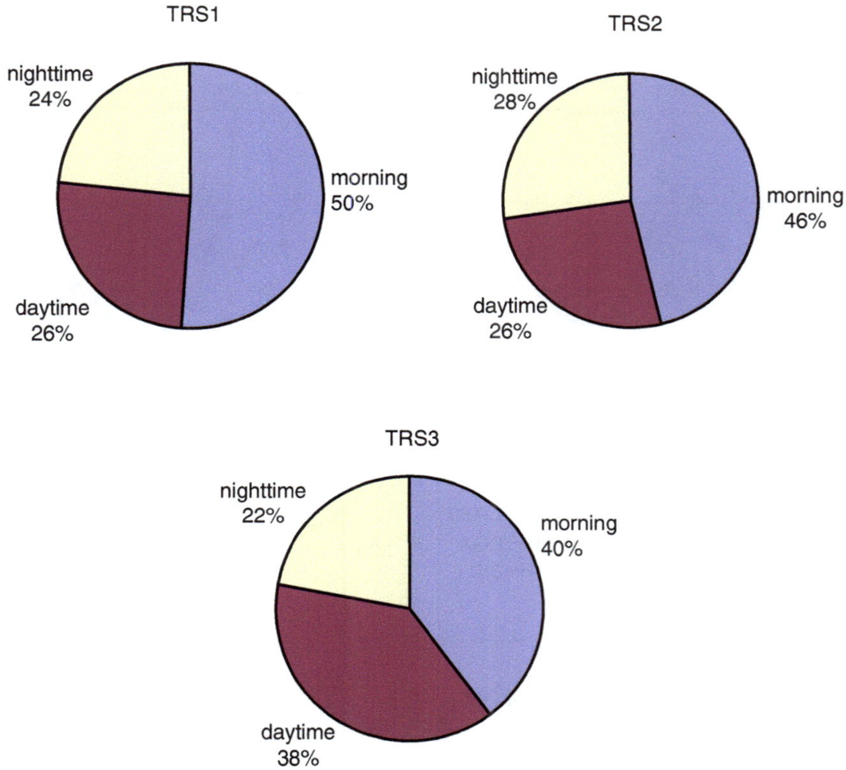

Fig. 5-17. Percentage contribution of additional cancer risk during a day in Tokyo

5.3.3.2 Risk Estimation of Potential Human Exposure to pPAHs in Bangkok

Similarly, the same investigation of lifetime cancer risk as described above was applied to the results in Bangkok. The monitoring results of Bangkok at all sampling sites in March and in August 2001 were calculated as the hourly average of pPAHs concentrations of weekdays and weekends, and of the whole sampling period. The hourly average profiles are shown in Figs. 5-18 and 5-19, respectively. In addition, the exposure of pedestrians and motorized road users to pPAHs in Bangkok during the period of 2:00–4:00 p.m. of August 10, 2001, and from 7:30 to 8:00 a.m. of August 14, 2001, were investigated. The pPAHs concentrations of both observations are displayed in Fig. 5-20.

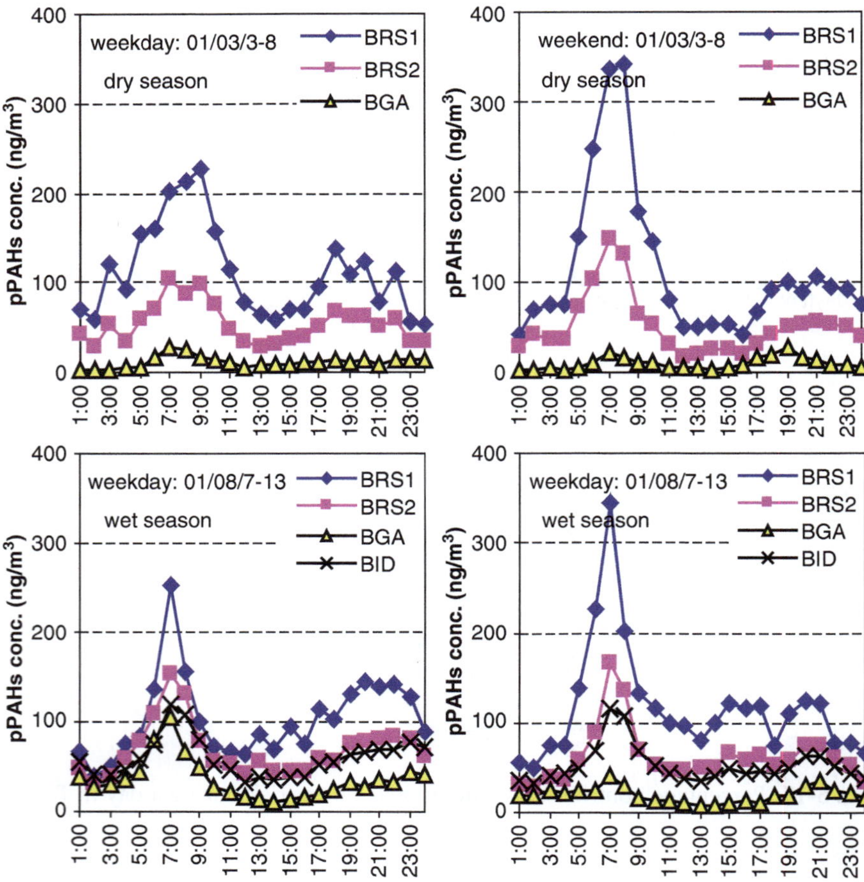

Fig. 5-18. Profile of hourly average pPAHs concentration of weekday and weekend during dry and wet season in Bangkok

To investigate the lifetime cancer risk, the total PEF-weighted concentration as expressed in the (5.12) was utilized, and the cancer risk is estimated by;

$$\text{Cancer risk} = \left(0.0091 \frac{\int_{t_1}^{t_2} C dt}{t_2 - t_1} + 0.2944 \right) \times 0.001 \times \frac{T_{\exp}}{24} \times 1.1 \times 10^{-3} \qquad (5.15)$$

As the BGA site represented the general area in this study, the cancer risk of the people at this site was then considered as background cancer risk.

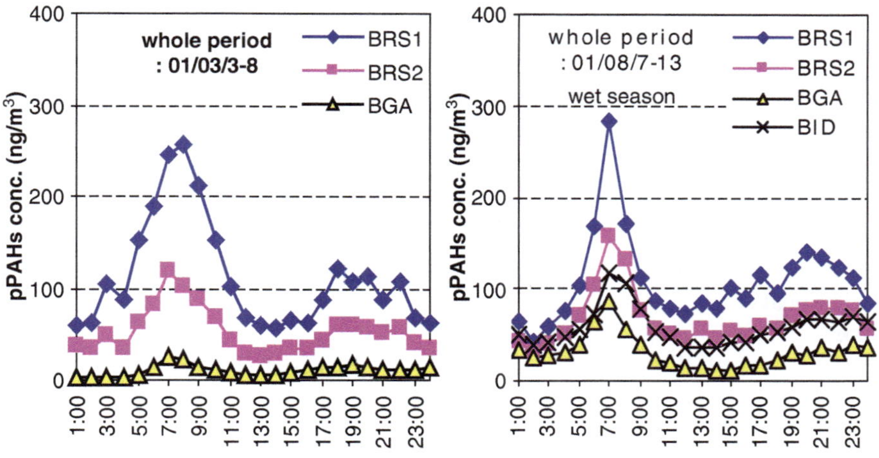

Fig. 5-19. Profile of hourly average pPAHs concentration of the whole sampling periods during dry and wet season in Bangkok

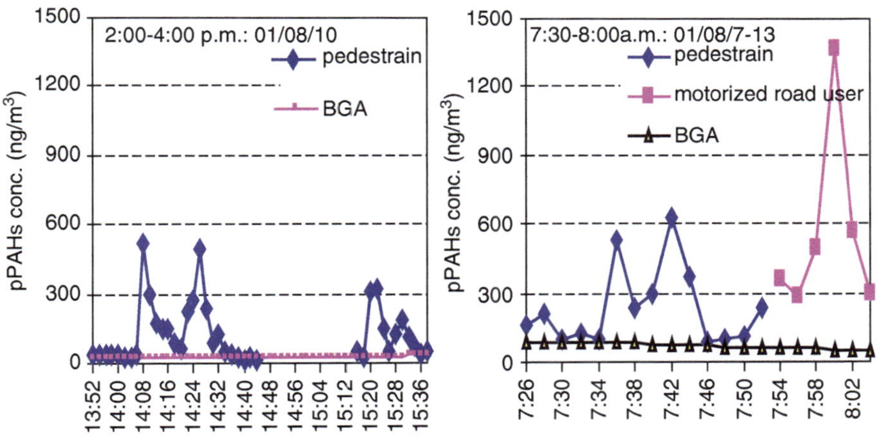

Fig. 5-20. pPAHs concentration observed for pedestrians and motorized road user in Bangkok

Consequently, the cancer risk of potential human exposure to pPAHs at the BRS1 and BRS2 site were considered as additional lifetime cancer risk due to road traffic. The results are summarized in Table 5-10.

The background cancer risk was estimated as 6.4×10^{-7} at the general area, the BGA site, in the second period, showing relatively higher than that of 4.3×10^{-7} in the first period. This is probably due to higher wind speed observed at this site, resulting in wider dispersion of the pPAHs during the

Table 5-10. Background and additional lifetime cancer risk of potential human exposure in the vicinity of road in Bangkok

	March			August		
	Whole period	Weekday	Weekend	Whole period	Weekday	Weekend
Background risk						
BGA	4.3E-07	4.3E-07	4.2E-07	6.4E-07	6.8E-07	5.3E-07
Additional risk						
BRS1	1.3E-06	1.3E-06	1.3E-06	1.1E-06	1.0E-06	1.3E-06
BRS2	7.4E-07	7.5E-07	7.3E-07	6.8E-07	6.5E-07	7.5E-07
BID				6.0 E-07	5.7E-07	6.6 E-07

second measurement. This indicates that the exposure of the people at the general area in the wet season was relatively higher than that in the hot season.

Comparing weekday and weekend, the results in Bangkok gave different from those observed in Tokyo. No difference in either background or additional risk at all sampling sites was observed in Bangkok, while the cancer risk during on weekend was almost half of that on weekday in Tokyo. Average value of the additional cancer risk at roadside due to the road traffic was calculated as 1.2×10^{-6} and 7.1×10^{-7} at ground (BSR1) and four-storey height levels (BRS2), respectively. The percentage decrease of the additional cancer risk from the ground level to the four-storey height level was approximately 40%. At the same receptor level, no large difference in the additional risk between indoor and outdoor area was observed, giving the ratio of indoor to outdoor exposure risk of 0.88. For the results in Tables 5-9 and 5-10, both of the background and the additional cancer risk estimated in Bangkok were relatively higher than those in Tokyo.

From Fig. 5-19, during the whole sampling periods in Bangkok, the percentage contribution of the additional cancer risk during a day at the ground level in roadside area was also determined, and the results are illustrated in Fig. 5-21. The additional cancer risk during the morning period contributed 35–43% of the total additional cancer risk in a day. For daytime and nighttime periods, the corresponding values were 28–34%, and 29–32%, respectively. As the results, no large difference in the range of percentage contributions of the additional cancer risk was observed. This indicates that the potential personnel exposure to total pPAHs for the people in proximity of road in Bangkok was almost the same throughout a day.

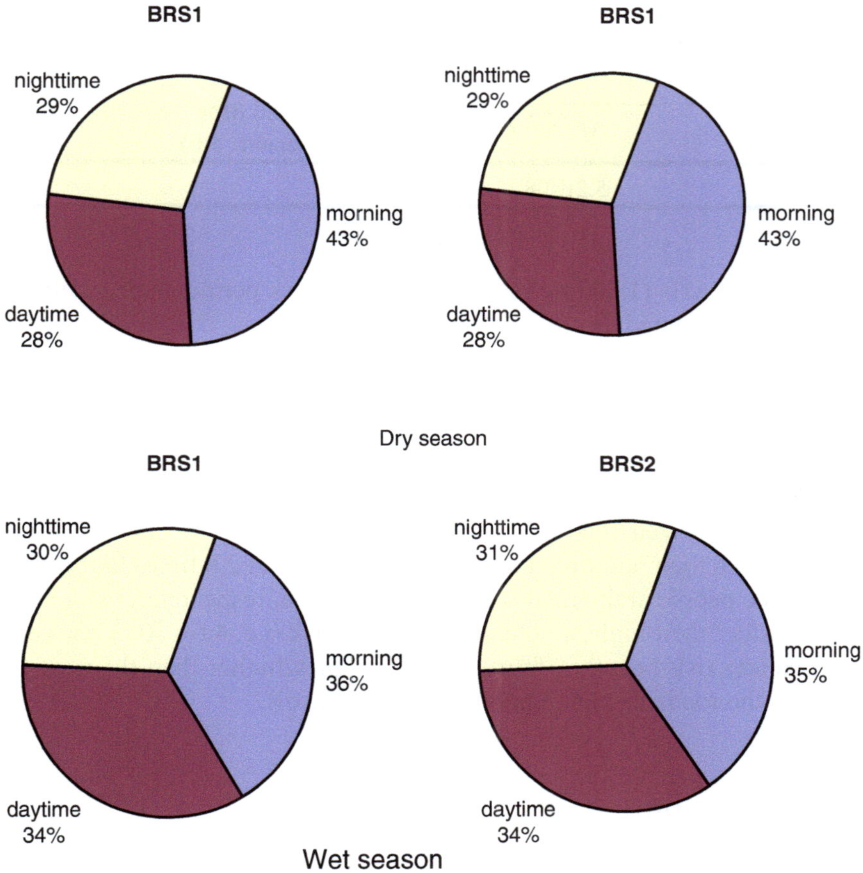

Fig. 5-21. Percentage contribution of additional cancer risk during a day at the roadside area during the whole sampling period in Bangkok

From Fig. 5-20, the additional risk of pedestrian and of motorized road users (i.e. motorbike or motor tricycle driver) was also estimated during the sampling period in wet season as shown in Table 5-11.

From Table 5-11, the additional cancer risk of pedestrian either in the morning or in the afternoon was similar. Then, if the people spent the time everyday for walking beside the main street for approximately 1 h, their additional cancer risk was estimated as 5.5×10^{-8}. In the case of motorized road user (i.e. motor tricycle drivers), the cancer risk was estimated as occupational group, assuming exposure time of 8 h workday. Consequently, the additional risk of the motor tricycle driver during workday was estimated as 1.8×10^{-6}. From Fig. 5-20, if the motorized road user exposure to the highest

Table 5-11. Additional lifetime cancer risk of pedestrian and motorized road user

Pedestrian (exposure period :1 h)		
Morning	Afternoon	Motorized road user (exposure period: 8 h)
5.9E-08	5.2E-08	1.8E-06

level of the pPAHs (1,359 ng/m^3) throughout a work period, their additional risk will be estimated as 4.6×10^{-6}.

According to the criteria for risk reduction under the California's Air Toxics Hot Spots Information and Assessment Act of 1987 (Collins and Brown 1998), all the estimated lifetime cancer risks in this study are still at a level not requiring public notification (since $< 10^{-5}$). However, we should carefully realize and pay attention to the potential personnel exposure to pPAHs of motorized road users, especially motor tricycle and motorbike drivers, which their additional cancer risk was 1.5 and 2.8 times higher than those of the people at the roadside and general area, respectively. Moreover, if we consider their highest potential exposure risk (i.e. 4.6×10^{-6}), the additional cancer risk becomes four and seven times higher than those of the people at the roadside and general area, respectively.

5.4 Conclusion

Advanced monitoring by a photoelectric aerosol sensor (PAS) could be utilized to clarify temporal and spatial variation of particle-bound PAHs. Further, this technique was successfully applied to assess the potential risk associated with human exposure to pPAHs by integrating with off-line technique and applying the potency equivalent factors (PEFs) scheme. The similar diurnal changes of pPAHs concentration were observed during the whole sampling period in Tokyo and in Bangkok. The pPAHs concentration increased in the early morning in accordance with a sudden burden of road traffic, and followed by a marked reduction in the daytime probably due to rising in the mixing zone caused by temperature increase in the daytime. In proximity of road either in Tokyo or in Bangkok, indoor pPAHs concentration with no obvious indoor source responded well to that at outdoor at the same receptor level with the certain time lags of 20 and 16 min, respectively, because the transportation of pPAHs from outdoor to indoor required a certain time. The spectral analysis clearly shows 1-, 0.5-, 7-, 1.2-, 2.3-, and 0.3-d periodicity of pPAHs for the urban sampling site especially at the roadside in Tokyo.

The quantitative differences in pPAHs concentrations were examined between Tokyo and Bangkok and there were differences among the sites in the level of the pPAHs concentration. The average pPAHs concentration at the roadside in Bangkok (53 ng/m^3) was relatively higher than that in Tokyo (29 ng/m^3); however, a bit higher concentration of the pPAHs at the general area in Tokyo (19 ng/m^3) than that in Bangkok (16 ng/m^3) was obtained. Differences observed either among the sites or between the cities in the pPAHs transportation patterns might be strongly affected by the surrounding sampling site's configuration and local wind characteristic in both cities.

Integrating on-line and off-line measurement could be successfully applied to assess the potential risk associated with human exposure to pPAHs. The percentage contribution of the additional cancer risk due to road traffic during the morning accounts for approximately 40–50% of the total additional risk in a day in Tokyo, while the contributions in the morning, daytime and nighttime in Bangkok were similar. The background and the additional cancer risk estimated in Bangkok were relatively higher than those in Tokyo. The background cancer risk at the general area was estimated as 3.3×10^{-7} for Tokyo, and 5.4×10^{-7} for Bangkok. In Tokyo, the additional cancer risk at the four-storey height in the roadside area was about 2.5×10^{-7} in summer. The additional cancer risk of 1.0×10^{-6} was observed at ground level in the roadside area in winter. In Bangkok, an additional cancer risk at roadside was estimated as 1.2×10^{-6} and 7.2×10^{-7} at ground and four-storey height levels, respectively. In the case of motorized road users, especially motorbike and motor tricycle drivers, their additional risk was estimated as 1.8×10^{-6}. As a result, the potential personnel exposure to pPAHs of the people living in the vicinity of road especially at the ground level and the road users would provide fundamental data for risk reduction in order to improve their quality of life.

References

Agnesod G, Maria RD, Fontana M, Zublena M (1996) Determination of PAH in airborne particulate: comparison between off-line sampling techniques and anautomatic analyzer based on a photoelectric aerosol sensor. Sci Total Environ 189/190:443–449

Ando M, Katagiri K, Tamura K, Yamamoto S, (1996) Indoor and outdoor air pollution in Tokyo and Beijing supercities. Atmos Environ 30:695–702

Collins JF, Brown JP, Alexeeff GV, Salmon AG (1998) Potency equivalency factors for some polycyclic aromatic hydrocarbons and polycyclic aromatic hydrocarbon derivatives. Regul Toxicol Pharmacol 28:45–54

Fischer PH, Hoek G, Reeuwijk H, Briggs DJ, Lebert E, Wijnen JH, Kingham S, Elliott PE (2000) Traffic-related differences in outdoor and indoor concentrations of particles and volatile organic compounds in Amsterdam. Atmos Environ 34:3713–3722

Hies T, Treffeisen R, Sebald L, Reimer E (2000) Spectral analysis of air pollutants. Part 1: elemental carbon time series. Atmos Environ 34:3495–3502

Kingham S, Briggs D, Elliott P, Fischer P, Lebert E (2000) Spatial variations in the concentrations of traffic-related pollutants in indoor and outdoor air in Huddersfield, England. Atmos Environ 34:905–916

McDow SR, Giger W, Burtscher H, Schmidt-Ott A, Siegmann HC (1990) Polycyclic aromatic hydrocarbons and combustion aerosol photoemission. Atmos Environ 24A:2911–2916

Nisbet IT, LaGoy PK (1992) Toxic equivalency factors (TEFs) for polycyclic aromatic hydrocarbons (PAHs). Regul Toxicol Pharmacol 16:290–300

OEHHA (1999) Air toxic hot spots program risk assessment guidelines: Part II Technical support document for describing available cancer potency factors, California, 100–110

Siegmann K, Siegmann HC (1998) Molecular precursor of soot and quantification of the associated health risk. Current Problems in Condensed Matter. Plenum, New York, pp 143–160

StatSoft (1999) STATISTICA for windows, vol III, USA

Nilrit S, Sathapanacharoo T, Prueksasit T (2005) Emission factors of polycyclic aromatic hydrocarbons from heavy and light duty diesel vehicles. Master thesis, Inter-department of Environmental Science, Chulalongkorn University, Bangkok, Thailand

Chetwittayachan T, Shimazaki D, Yamamoto K (2002a) A comparison of temporal variation of particle-bound polycyclic aromatic hydrocarbons (pPAHs) concentration in different urban environments: Tokyo, Japan, and Bangkok, Thailand. Atmos Environ 36:2027–2037

Chetwittayachan T, Shimazaki D, Yamamoto K (2002b) Integrating on-line and off-line measurement for assessment of potential human exposure to particle-bound polycyclic aromatic hydrocarbons (pPAHs) in Bangkok, Thailand. Tenth international conference on modeling, monitoring and management of air pollution, Air pollution X, WIT, Southampton, Boston, UK, 741–750

Chetwittayachan T, Kido R, Shimazaki D, Yamampoto K (2002c) Diurnal profiles of particle-bound polycyclic aromatic hydrocarbon (pPAH) concentration in urban environment in Tokyo metropolitan area. Water Air Soil Pollut Focus 2:203–221

US EPA (1999) Method-TO-13A-determination of polycyclic aromatic hydrocarbons (PAHs) in ambient air using gas chromatography/mass spectrometry (GC/MS). Compendium of methods for the determination of toxic organic compounds in ambient air, EPA/625/R-96/010b, USA

MIX
Papier aus verantwortungsvollen Quellen
Paper from responsible sources
FSC® C105338

If you have any concerns about our products,
you can contact us on
ProductSafety@springernature.com

In case Publisher is established outside the EU,
the EU authorized representative is:
Springer Nature Customer Service Center GmbH
Europaplatz 3, 69115 Heidelberg, Germany

Printed by Libri Plureos GmbH
in Hamburg, Germany